SCHWERTMANNITE

This book offers a comprehensive overview of the structure and solubility of the mineral schwertmannite, while also delving into characterization methods and techniques, environmental stability, and the potential to retain pollutants.

Using examples from the 1990s, when the mineral was first discovered, *Schwertmannite* compiles years of research in this supplementary guide. Chapter 1 explains its occurrence in various environmental settings, the conditions of its formation, and its association with other secondary precipitates. Chapter 2 discusses its solubility, structure, and iron–sulphur coordination, while Chapter 3 delves into synthesizing approaches such as dialysis techniques, oxidation pathways, and urea hydrolysis. Chapter 4 explores how schwertmannite came to be characterized based on colour, composition, morphology, and crystallinity. The last two chapters deal with the mineral's interaction with contaminants, its binding mechanisms, its metastability with respect to other iron oxyhydroxides, and the factors controlling transformation processes.

This book will benefit a large community of scientists and researchers working in acid mine drainage, since schwertmannite is the most abundant mineral in these sulphate-rich environments. It will be a valuable body of knowledge for research organizations and mining companies, as well as those teaching courses on mining, mineral chemistry, wastewater treatment, acid sulphate soils, soil geochemistry, and mine reclamation.

SCHWERTMANNITE

Formation, Stability and Applications

Susanta Paikaray and Stefan Peiffer

CRC Press is an imprint of the
Taylor & Francis Group, an **informa** business

Designed cover image: A SEM image of the schwertmannite that was synthesized through bacterial oxidation of Fe2+from SO4 2--rich lignite mine effluents using Acidithiobacilius ferrooxidan between pH 2.9 and 3.2 at GEOS, Freiberg, Germany (© Elsevier)

First edition published 2026
by CRC Press
4 Park Square, Milton Park, Abingdon, Oxon, OX14 4RN

and by CRC Press
2385 NW Executive Center Drive, Suite 320, Boca Raton FL 33431

CRC Press is an imprint of the Taylor & Francis Group, an informa business

British Library Cataloguing-in-Publication Data
A catalogue record for this book is available from the British Library

ISBN: 9781032952406 (hbk)
ISBN: 9781032952383 (pbk)
ISBN: 9781003583875 (ebk)

DOI: 10.1201/9781003583875

Typeset in Sabon
by codeMantra

CONTENTS

PREFACE

The number of water bodies affected by acid mine drainage as a result of the oxidation of pyrite during mining activities is huge. For the USA alone, it is estimated that watercourses with a total length of around 20,000 km are affected. A similar situation can be expected in many other countries, and the proportion of affected regions and water bodies is likely to increase in the future. One of the most common side effects of acid mine drainage is the precipitation of Fe(III)- and sulphate-rich minerals, of which Schwertmannite is the most commonly occurring mineral. It occurs ubiquitously in areas affected by acid mine drainage, especially after the highly concentrated pyrite oxidation products have diluted somewhat. The mineral is an important geochemical buffer whose presence adjusts the pH of surface waters affected by acid mine drainage to characteristic values between 2.5 and 4. Schwertmannite is known for its high uptake capacity for all possible contaminants typically found in acid mine drainage. The mineral therefore represents an important sink function for pollutants. These properties have therefore made schwertmannite the focus of scientific research with regard to its use as a sorbent material for the purification of various wastewater streams. Yet, a comprehensive text book that brings together the various facets about this mineral ranging from mineral properties and synthesis to occurrence and relevance in the environment is still missing despite the existence of excellent review articles. Due to this enormous importance, we have decided to write this book *Schwertmannite—Formation, Stability and Application*.

This book aims to provide the reader with detailed knowledge of the various aspects of the mineral schwertmannite, such as the solubility question, characterization methods and techniques, environmental stability, and its potential to retain contaminants. The current state of research knowledge has been compiled with the aim of providing information on all relevant aspects in one place. Chapter 1 discusses the occurrence of schwertmannite in different environments, its formation conditions, and its association with other secondary precipitates, while Chapter 2 deals with its solubility, structure, and iron–sulphur coordination. Chapter 3 discusses various synthesis processes, such as dialysis or oxidation processes. Chapter 4 deals with techniques and methods for the characterization of schwertmannite. The interactions with impurities and the binding mechanisms are explained in Chapter 5, while its metastability with respect to other iron oxyhydroxides and the factors controlling the transformation processes are discussed in Chapter 6.

We expect this book to benefit a wide readership engaged in research, application, and teaching related to acid mine drainage. Courses in mining, mineral chemistry, wastewater treatment, acid sulphate soils, soil geochemistry, and mine reclamation will greatly benefit from this book, and it will be a valuable body of knowledge for research institutions and mining companies.

Susanta Paikaray and Stefan Peiffer, May 2025

ACKNOWLEDGEMENTS

The knowledge compiled in this book is based on many years of research by the two authors in the field of acid mine drainage, which was made possible by numerous research grants. We are grateful to the funding we obtained over the years from the German academic exchange service (DAAD), the Deutsche Forschungsgemeinschaft (DFG), the German Ministry of Science and Research (Geotechnologien Programme), and others. The University Grants Commission (UGC) India is acknowledged for technical and infrastructural support while preparing this book.

We greatly acknowledge the contributions from all who came across the work paths on schwertmannite by sharing their knowledge and contributing to our understanding on this mineral. To name a few, Christian Blodau, Jan Fleckenstein, Klaus H Knorr, Markus Bauer, Julia Beer, Katrin Hellige, Jutta Eckert, Silke Hammer, Martina Rohr, and others have helped a lot during the work on schwertmannite.

AUTHOR BIOGRAPHIES

Susanta Paikaray is presently working on various aspects of aquatic geochemistry along alluvial plains. He currently works as an Assistant Professor at Panjab University Chandigarh (India) in the field of Environmental Geology, Climatology, and Aqueous Geochemistry. He obtained his PhD degree on environmental and sedimentary geochemistry from IIT Bombay (India) and MSc in Applied Geology from IIT Roorkee (India). His research interests include mine drainage, wastewater treatment, mineral and sediment geochemistry, hydrochemistry and investigating soil, groundwater and crop pollution by metallic and non-metallic contaminants, microplastics in aquatic environment, mine tailing, and urban waste management. He has been working on schwertmannite for several years, especially aspects of stability, chemistry, and environmental applications.

Stefan Peiffer is working on the geochemistry of aquatic systems for more than 30 years with a focus on coupled geochemical and microbial processes in sedimentary and groundwater environments. With a special emphasis on the interaction between the iron and sulphur cycle, his present research includes the kinetics and mechanisms of surface-mediated mineral transformations. He has also studied the effect of advective groundwater flow on biogeochemical process rates during groundwater–surface water interactions, with attention to acid mine drainage affected systems including schwertmannite as a key mineral.

1

SCHWERTMANNITE OCCURRENCES

1.1 Introduction

Atmospheric exposure of iron sulphide minerals, such as pyrite (FeS_2) and arsenopyrite (FeAsS), around mine dumps and waste piles releases large amounts of SO_4^{2-} and Fe^{2+} into nearby aquatic environments, such as streams and water channels. This leads to severe acidification of the affected water bodies, with pH levels as low as 1.0. Such acidity is a major global concern because it causes the dissolution of nearly all mineral phases. In addition to iron sulphides, similar acidic conditions can develop around other sulphide mines, such as those containing lead (PbS), zinc (ZnS), copper (CuS), and nickel (NiS). Toxic metals, including As, Pb, Cd, Zn, Cu, and Ni, are mobilized due to the dissolution of gangue materials in these extremely acidic environments, resulting in their enrichment in the surrounding ecosystem.

The Fe^{2+} released during the oxidative dissolution of FeS_2 (Eq. 1.1) is simultaneously oxidized to Fe^{3+} by atmospheric O_2 (Eq. 1.2) and/or by iron-oxidizing bacteria (discussed below). The Fe^{3+} produced in this process accelerates FeS_2 oxidation in some cases and promotes ferric hydroxide precipitation through hydrolysis in others (Eq. 1.3). The reaction kinetics show that the generation of acidity is eight orders of magnitude greater when sulphides are oxidized by Fe^{3+} (Eq. 1.4) compared to oxidation by atmospheric O_2 (Eq. 1.1). Since oxidation by atmospheric O_2 is extremely slow, Fe^{2+} generation is considered the rate-determining step in the dissolution of iron sulphides (Evangelou and Zhang, 1995).

$$FeS_2 + 3.5O_2 + H_2O \rightarrow Fe^{2+} + 2H^+ + 2SO_4^{2-} \qquad \text{(Eq. 1.1)}$$

DOI: 10.1201/9781003583875-1

$$Fe^{2+} + 0.25O_2 + H^+ \rightarrow Fe^{3+} + 0.5H_2O \quad \text{(Eq. 1.2)}$$

$$Fe^{3+} + 2.5H_2O + 0.25O_2 \rightarrow Fe(OH)_3 + 2H^+ \quad \text{(Eq. 1.3)}$$

$$FeS_2 + 14Fe^{3+} + 8H_2O \rightarrow 15Fe^{2+} + 2SO_4^{2-} + 16H^+ \quad \text{(Eq. 1.4)}$$

In the presence of acidophilic iron-oxidizing bacteria, such as *Thiobacillus* species, the oxidation rate of Fe^{2+} can be accelerated five to six times compared to abiotic conditions (Singer and Stumm, 1970; Nordstrom, 1982; Dold, 2014). Marcasite, an orthorhombic dimorph of pyrite, also produces the same amount of acid as pyrite, given its identical formula (FeS_2). The ferric hydroxide produced by Fe^{2+} oxidation (Eq. 1.3) and Fe^{3+} hydrolysis (Eq. 1.3) is not always pure, rather a mixture of iron-bearing minerals, including ferrihydrite, schwertmannite, lepidocrocite, and goethite.

The precipitates formed along mine drainage channels, especially under acidic pH conditions, are often reported as "amorphous ferric hydroxides" without specific identification. These amorphous precipitates contain a high content of SO_4^{2-}, differing significantly from other minerals commonly found in the same environments, such as jarosite and goethite. Thermodynamic modelling of mine drainage data indicated equilibrium with SO_4-bearing ferric hydroxide phases that show solubility around pH ~ 3 (Karathanasis et al., 1988; Sullivan et al., 1988).

Due to its amorphous nature, this mineral phase lacked proper identification until the 1990s when Bigham et al. (1990, 1994) systematically characterized it and named it schwertmannite, in honour of Prof. Dr. Udo Schwertmann from the Technical University of Munich, Germany (1927–2016), for his significant contributions to the study of iron minerals. Schwertmannite is commonly found in acidic mine environments, where heavy metal pollutants are discharged due to the leaching of waste piles. The behaviour of ions and trace metals in these environments influences the fate of schwertmannite and the contaminants it interacts with.

Thus, a brief overview of acid mine drainage (AMD) is provided, followed by a discussion of various secondary precipitates observed in mine drainage areas, and schwertmannite's occurrence beyond these environments.

1.2 AMD and secondary iron minerals

1.2.1 Acid mine drainage

AMD occurs when acid production from pyrite oxidation (Eqs. 1.1 and 1.4) and ferric iron precipitation (Eq. 1.3) exceeds the rate of acid neutralization due to the solution's alkalinity and reactions with surrounding minerals.

The mechanisms of acidity generation in AMD regions remain a subject of ongoing research and debate, but are recognized as a high-priority issue globally (Nordstrom, 2011). Although the specific details vary across mine regions, a general understanding involves pyrite (FeS_2) releasing high concentrations of Fe^{2+} and SO_4^{2-}, which contribute to the acidity of mine waters (Eq. 1.1). Pyrite is highly insoluble but rapidly oxidizes in the presence of atmospheric O_2 and microbes with iron-oxidizing bacteria like *Acidithiobacillus ferrooxidans* and *Leptospirillum ferrooxidans* playing a crucial role in acidity generation (Baker and Banfield, 2003) in mediating Eq. (1.2), thereby generating Fe^{3+}. Since Fe^{3+} is a stronger oxidant than O_2, the presence of these microbes accelerates the process. Both direct and indirect mechanisms govern this bioleaching. In the direct mechanism, bacteria attach to sulphide minerals, facilitating enzymatic attack, while bacterially produced Fe^{3+} chemically dissolves the sulphide mineral in the indirect mechanism.

The role of bacteria is to regenerate Fe^{3+} and H^+, concentrating them at the mineral surface to enhance dissolution. Thiosulphate is the first intermediate in FeS_2 oxidation, spontaneously oxidizing to sulphate through intermediate species such as tetrathionate and trithionate (Schippers et al., 1996; Küsel, 2003). Fe^{3+} generated from Fe^{2+} catalyzes further FeS_2 oxidation, producing Fe^{2+}, which is then cyclically oxidized back to Fe^{3+} until all FeS_2 is consumed. Electron transfer between Fe^{2+} and Fe^{3+} involves one electron, while sulphur transfer involves seven electrons, forming intermediate sulfoxyanions that quickly transform into sulphate.

The acidity of AMD is not only determined by the concentration of H^+ ions, but also by the presence of Fe^{3+} (Eq. 1.3) and Al^{3+} (Eq. 1.5). The hydrolysis of Fe^{3+} and Al^{3+} (at pH ~ 2.7 and ~4.5, respectively) releases significant additional acidity, contributing to the buffering capacity of AMD systems.

$$Al^{3+} + 3H_2O \rightarrow Al(OH)_3 + 3H^+ \qquad \text{(Eq. 1.5)}$$

Various sulphide minerals are present in mine tailing impoundments, each with a specific susceptibility to oxidation. In addition to FeS_2, sulphide minerals such as galena (PbS), sphalerite (ZnS), chalcopyrite ($CuFeS_2$), and arsenopyrite (FeAsS) also contribute significantly to mine drainage acidity (Jamieson et al., 2015). The oxidation rates of these minerals by O_2 and Fe^{3+} vary considerably depending on their reactivity. Lindsay et al. (2015) proposed the general oxidation trend for common sulphide minerals as follows: pyrrhotite ($Fe_{1-x}S$) > galena (PbS) > sphalerite ($Zn_{1-x}Fe_xS$) > bornite (Cu_5FeS_4) > pentlandite $(Fe,Ni)_9S_8$) > arsenopyrite (FeAsS) > marcasite

(FeS_2) > pyrite (FeS_2) > chalcopyrite ($CuFeS_2$) > magnetite ($(Fe^{2+}Fe^{3+})_2O_4$) > molybdenite (MoS_2). This sequence is based on optical observations of thin sections from various impoundments. Although magnetite is not a sulphide, it is included in this list due to its frequent occurrence alongside sulphides in mine wastes.

1.2.2 Involvement of microbes in AMD generation

AMD localities are often considered biologically dead due to their high acidity, but these environments are not devoid of life. Various forms of biota, including bacteria, algae, archaea, yeast, and fungi, thrive in such conditions. The presence of oxidizable dissolved iron and reduced sulphur compounds, common in mine drainage environments, promotes the growth of chemolithoautotrophic bacteria. These bacteria, classified as "autotrophic", obtain their carbon from dissolved or atmospheric CO_2, while deriving energy from the oxidation of inorganic sources like reduced Fe and S, making them "chemolithotrophic".

Once colonies of chemoautotrophs form, their degradation products provide organic carbon for heterotrophic bacteria, which derive energy from organic compounds, as well as for eukaryotes. Detailed studies of biofilms from AMD-affected areas have revealed a wide variety of bacterial communities. For example, research at Iron Mountain, California, identified a dominance of *Leptospirillum rubarum*, *Leptospirillum ferrodiazotrophum*, and *Ferroplasma acidarmanus* (Edwards et al., 2000). Other less common species found include *Sulfobacillus*, *Acidimicrobium*, *Acidiphilium*, and *Acidithiobacillus* (Bond et al., 2000).

Sulphur-oxidizing bacteria commonly found in sulphide tailings include neutrophiles (e.g., *Thiobacillus thioparus*) and acidophiles (e.g., *Acidithiobacillus thiooxidans*), which thrive under neutral mine drainage (NMD) and AMD conditions, respectively. These bacteria cannot oxidize sulphide minerals directly but gain energy by oxidizing S^{2-}, S^0, $S_2O_3^{2-}$, and other intermediate sulphur species. Complete oxidation generates H^+ and SO_4^{2-}, limiting the accumulation of intermediate sulphur species produced by sulphide-mineral oxidation. The generation of H^+ also promotes the non-oxidative dissolution of acid-soluble sulphide minerals.

Similarly, iron-oxidizing bacteria obtain energy by oxidizing Fe^{2+} to Fe^{3+}. Microbial Fe^{2+} oxidation rates in acidic environments are up to 10^5times faster than abiotic rates (Nordstrom and Southam, 1997; Nordstrom, 2003).

Geobacter and *Shewanella* species are thought to play important roles in Fe^{3+} reduction in sedimentary environments. In acidic mine environments, the reduction of Fe^{3+} coupled with the oxidation of organic matter

(OM) appears to be mediated by acidophilic *Acidiphilium* species. *Thiobacillus ferrooxidans* is the predominant iron-oxidizing bacterium in AMD localities, capable of oxidizing reduced sulphur while reducing Fe^{3+} under anaerobic conditions (Küsel and Dorsch, 2000). Other iron/sulphur-oxidizing bacteria, such as *L. ferrooxidans*, *Thiobacillus thiooxidans*, and *Sulfobacillus thermosulfidooxidans*, are also found in AMD waters. Kim et al. (2002) reported the presence of *Gallionella ferruginea*, associated with brownish-yellow schwertmannite precipitates, in the Donghae mine, South Korea.

1.2.3 Secondary precipitates of iron

Minerals associated with AMD are categorized into four groups: primary, secondary, tertiary, and quaternary minerals (Amos et al., 2015; Nordstrom et al., 2015). Primary minerals are those originally present in the ore and gangue minerals, including sulphide ores, carbonates, and silicates. Secondary minerals form in mine wastes and include compounds such as gypsum, jarosite, and goethite. Tertiary minerals develop after samples are removed from mine workings and include minerals like calcium, magnesium, and iron(II) sulphates. Quaternary minerals, which form during the storage of dried samples, are typically hydrated iron sulphates such as rozenite ($FeSO_4{\cdot}4H_2O$) or siderotil ($FeSO_4{\cdot}5H_2O$).

AMD effluents are naturally neutralized after exiting mine adits. This neutralization is driven by the dissolution of hydroxyl-releasing minerals, such as aluminosilicates, clays, and carbonates, which buffer the acidity until these minerals are depleted. As this process continues, the pH of the drainage increases from acidic (pH 2.5–3.5) to alkaline conditions (pH > 6.5–7.0). During this neutralization, dissolved ions in AMD effluents, including Fe^{3+}, Al^{3+}, Mg^{2+}, Ca^{2+}, SO_4^{2-}, and $CO_{2(aq)}$, gradually precipitate as secondary mineral phases. The formation of these secondary precipitates is governed by their solubility product constant (pK_{sp}), with minerals having lower pK_{sp} values forming earlier as oxides, hydroxides, hydroxysulphates, or carbonates. Dzombak and Morel (1990) suggested that the precipitation of hydrolysable metal ions, such as Fe^{3+}, typically occurs near their pK_{sp} values. Research on various mine effluents (e.g., Lee et al., 2002) has demonstrated that Fe-rich precipitates form in a sequence governed by the increase in pH. Beginning at pH < 2, jarosite is the first mineral to form, followed by schwertmannite in the pH range ~ 2.5–4.5 and ferrihydrite (pH ~ 5–6).

Under acidic conditions, Fe^{2+} remains stable unless oxidized by prolonged atmospheric interaction or bacterial metabolism, leading to the formation of ferrous sulphate minerals such as melanterite ($FeSO_4{\cdot}7H_2O$), ferrohexahydrite ($FeSO_4 \cdot 6H_2O$), siderotil ($FeSO_4 \cdot 5H_2O$), rozenite ($FeSO_4 \cdot 4H_2O$), and szomolnokite ($FeSO_4 \cdot H_2O$) (Murad and Rojík, 2004). The oxidation of Fe^{2+}, either by atmospheric O_2 and/or bacterial processes, is favoured at pH levels between 5 and 9. Along with oxidation, the parallel hydrolysis of Fe^{3+} leads to the formation of various ferric hydroxides and oxyhydroxysulphates, such as jarosite ($MFe_3(SO_4)_2(OH)_6$, where M = K^+, Na^+, NH_4^+, H_3O^+ etc.), schwertmannite ($Fe_8O_8(OH)_6SO_4$), goethite (α-FeOOH), and ferrihydrite ($Fe_2O_3 \cdot 0.5H_2O$).

Typically, five different Fe-containing secondary precipitates are found in AMD environments: jarosite, schwertmannite, goethite, ferrihydrite, and amorphous precipitates (Marescotti et al., 2012). From these mineral phases, jarosite and goethite very closely co-occur with schwertmannite. These precipitates exhibit distinct colours that aid in visual identification. For instance, jarosite-rich precipitates appear yellowish-red, schwertmannite-rich ones are yellowish-brown, and goethite-rich precipitates are brownish-yellow. These correspond to Munsell colours 5YR 5/8, 10YR 5/8, and 10YR 6/6, respectively (Table 1.1, Murad and Rojík, 2004).

Individual precipitates in AMD environments are associated with characteristic pH, redox potential (Eh), and SO_4^{2-} concentrations. Jarosite-dominated precipitates typically form at the lowest pH (1.5–3.0), the highest Eh (~620 mV), and the highest SO_4^{2-} (>3000 mg L^{-1}) conditions. In these environments, concentrations of Fe, Al, and dissolved metals such as Zn, Cu, Co, Mn, and As are elevated due to the strong acidity. Jarosite types, including K-jarosite, Na-jarosite, and NH_4-jarosite, depend on the dominant cations present in the aqueous phase. Minor phases, such as goethite, quartz, and clay minerals, are often present. Schwertmannite dominates at slightly higher pH (~3–4), with a lower Eh (560–590 mV), and moderate SO_4^{2-} levels (1000–3000 mg L^{-1}). In these conditions, trace metal concentrations remain high due to persistent acidity. Goethite-dominated

TABLE 1.1 Properties of ferric oxyhydroxides and oxyhydroxysulphate minerals formed around acid mine drainage–affected regions (ref. Murad and Rojík, 2004)

Mineral	*Munsell colour*	*XRD peaks*		*SSA (m^2 g^{-1})*
		n	*FWHM*	
Goethite	7.5YR–10YR	~25	0.5–1	100–200
Schwertmannite	10YR–2.5Y	8	3–7	125–225
Jarosite	2.5Y–5Y	~17	<0.1	Small

FWHM (full width at half maximum), °2θ Co K_α; n, no. of peaks for d > 1.5Å; SSA, specific surface area; XRD, X-ray diffraction.

precipitates form at relatively higher pH values (>3.7) and are associated with lower SO_4^{2-} concentrations (<1000 mg L^{-1}) and Eh compared to schwertmannite and jarosite (Bigham and Nordstrom, 2000; Murad and Rojík, 2003, 2004; Jönsson et al., 2005; Dakos et al., 2012). Goethite can form directly or through schwertmannite transformation, though this is less likely in highly acidic conditions. Another associated mineral phase lepidocrocite (γ-FeOOH) formation occurs at pH 7 upon Fe^{2+} oxidation by atmospheric O_2 in AMD effluents, and a mixture of lepidocrocite, goethite, ferrihydrite, and schwertmannite can form at pH 5.5 during the oxidation and aging of schwertmannite (Jönsson et al., 2006).

Schwertmannite typically dominates at pH ~ 3.8 with SO_4^{2-} concentrations >1500 mg L^{-1}, while goethite–schwertmannite mixtures occur at pH ~ 2.7, as observed in the Libiola Mine, Italy (Dinelli and Tateo, 2002). Spectral analysis by Williams et al. (2002) identified iron minerals such as schwertmannite and goethite in acidic (pH 3.1–5.8) coal mine drainage in eastern Pennsylvania, USA. Secondary minerals like schwertmannite (pH ~ 3.0), ferrihydrite, and goethite (pH > 5) form when Paroistenjärvi Mine effluent in Ylöjärvi, Finland mix with surface water (Carlson et al., 2002). In the vicinity of discharge points (pH 2–4), schwertmannite, jarosite, and goethite precipitates dominate. In Essouk River, Algeria, Boukhalfa and Chaguer (2012) found jarosite (pH < 2.2) and schwertmannite with minor goethite (pH > 3.2). Kim et al. (2002) identified three types of ochreous precipitates in the Donghae Mine, South Korea: reddish-brown, brownish-yellow, and white biomats, each associated with specific pH conditions. White biomats form at pH 4.4–5.6, while brownish-yellow precipitates occur at pH values below 4.0. Precipitates with pH between 4.0 and 4.4 exhibit a yellowish colour due to the mixing of the white and brownish-yellow biomats. The reddish-brown biomat consists mainly of ferrihydrite with minor goethite; the brownish-yellow biomat contains primarily schwertmannite with minor quartz, pyrophyllite, and illite; and the white biomats are composed of amorphous aluminium sulphates.

The precipitation of Fe(III) (oxy)hydroxides and hydroxysulphates typically occurs within the oxidation zone, while hydrated Fe(II) sulphates precipitate in water-unsaturated tailings below the oxidation zone. Extensive secondary mineral formation can create cemented layers of primary and secondary minerals, known as hardpan, which can limit pore-water movement and oxygen ingress, thereby inhibiting sulphide-mineral oxidation in the underlying tailings. Hardpans also serve as temporary sinks for metal(loid)s released during sulphide weathering. Ferric hydroxides and hydroxysulphates such as ferrihydrite, jarosite, lepidocrocite, and goethite dominate in the oxidized hardpan, while ferrous sulphates, including melanterite and rozenite, dominate below the oxidation zone. Fe(II) sulphate minerals typically form in the vadose zone, beneath the sulphide oxidation zone but above the water table. These efflorescent minerals, which form due to evaporation of AMD from tailings, waste rock, or mine workings, can also form hardpan layers along drainage paths (Lindsay et al., 2015). These minerals also have the potential to adsorb trace metal pollutants such as Zn, Cu, and Ni (Moncur et al., 2005).

1.3 The geochemical window for the occurrence of schwertmannite

Ferric precipitates formed by hydrolysis of ferric ions in acidic environments are often referred to as "yellow boy" due to their yellow colour, or "amorphous ferric hydroxides", though this does not include ferrihydrite. Schwertmannite typically forms in acid sulphate-rich environments, where rapid oxidation of Fe^{2+} occurs, especially within a pH range of 3.0 to 4.5 and sulphate concentrations between 1000 and 3000 mg L^{-1}. Schwertmannite tends to precipitate around pH 2.5–4.5, but can transform into jarosite or ferrihydrite/goethite outside this range. Some studies have even reported its formation in more acidic conditions, such as at pH ~ 2.2 (España et al., 2006) and pH ~ 2.0–3.4 in environments rich in *A. ferrooxidans*. A comprehensive study by Caraballo et al. (2013), which compiled data from 30 schwertmannite samples from different countries, suggests that schwertmannite can precipitate across a pH range from 1.93 to 4.71 and a pE (a measure of the oxidation potential) ranging from 8.50 to 13.70, with an average pH of 3.02 ± 0.56 and a mean pE of 11.37 ± 1.37. This demonstrates that schwertmannite is most commonly found in acidic, moderately oxidizing environments. At lower pH, jarosite may form if sulphate concentrations are high, and goethite, a common associated phase, may form as an alteration product of schwertmannite, particularly at pH > 4.5 (Sidenko and Sherriff, 2005).

Lazaroff et al. (1982) suggested that an amorphous ferric hydroxysulphate precipitate forms during the bacterial oxidation of $FeSO_4$ in acidic media by *T. ferrooxidans*, but this phase had no definitive name. The reddish-brown precipitate had an Fe/SO_4 ratio of 3.5–5.0. Earlier studies by Lazaroff (1963) emphasized the need for excess sulphate for bacterial oxidation of Fe^{2+}, which leads to the precipitation of ferric compounds distinct from goethite, ferrihydrite, and jarosite. This is consistent with Brady et al. (1986), who identified this phase as "amorphous Fe-oxides" in a stream affected by AMD in SE Ohio, USA. The presence of goethite next to amorphous ferric precipitate was evident due to the crystalline peaks observed in X-ray diffraction (XRD) analyses.

Brady et al. (1986) were among the first to report schwertmannite, originally describing it as an unnamed amorphous ferric sulphatic mineral phase with a yellowish colour, formed in low-pH drainage environments (Figure 1.1). It was later named schwertmannite after its detailed identification by Bigham et al. (1990, 1994) and has since been widely recognized in mine drainage environments. Because of its non-crystalline nature, schwertmannite has often been referred to in research reports as "amorphous ferric hydroxide".

Schwertmannite has been widely reported across multiple continents, including North and South America, Europe, Australia, and Asia. It frequently occurs in AMD (Bigham et al., 1990, 1994, 1996b; España et al., 2005), mine pit lakes (Peine et al., 2000; Regenspurg et al., 2004;

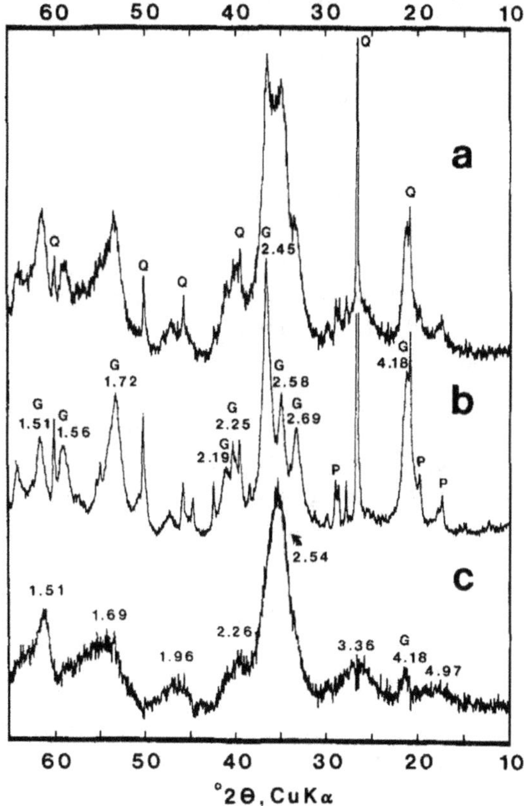

FIGURE 1.1 X-ray diffractograms of precipitates from AMD, Ohio, USA, of the original precipitate (a), after 15-minute oxalate dissolution (b), and after spectral subtraction of (b) from (a). G-goethite, Q-quartz, P-phyllosilicates (ref. Brady et al., 1986).

Blodau, 2006), natural streams (Schwertmann et al., 1995), and acid sulphate soils (Burton et al., 2006, 2007). More recently, it has even been identified on Mars (Bishop et al., 2004) and in Antarctica (Dold et al., 2013), demonstrating its potential interplanetary importance. Type specimens are now held in the collection of the Geological Museum at the University of Helsinki, Finland.

Although schwertmannite was accepted as a mineral by the Commission on New Minerals and Mineral Names in 1992 (90–006), its status as a mineral was later questioned by French et al. (2012). In mineralogy, a mineral is defined as a naturally occurring substance produced through geochemical or biogeochemical processes, possessing a highly ordered, repeating atomic arrangement (French et al., 2012). Its composition can be described by a fixed or variable chemical formula (French et al., 2012). A true mineral must exhibit consistent physical and chemical properties, including major and minor elements. This definition also encompasses nano-minerals and mineral nanoparticles, which may have slightly different physical properties due to their small size. Following this definition, schwertmannite, with its needle-like and whisker structures, was classified by French et al. (2012) as a nano-mineral.

However, discrepancies in schwertmannite's structure challenge its classification as a mineral. Its atomic structure includes ordered regions that are coherent only over a few to several nanometres. Some regions display d-spacing patterns similar to goethite and jarosite, while others resemble schwertmannite. Additionally, while the overall Fe and S content matches schwertmannite stoichiometry, the S/Fe ratio varies considerably along the edges of its needle-like structures, with some areas showing enrichment in silicon. Given these inconsistencies, French et al. (2012) questioned whether schwertmannite should be classified as a mineral or better as a mineraloid.

1.4 The formation of schwertmannite-bearing environments

1.4.1 Acid mine drainage

In dilute mine drainage environments, two distinct acidic buffer zones are observed at pH levels of ~2.7 and ~4.5 that correspond to the pH buffering exerted by the solubility of schwertmannite and aluminium precipitates, respectively (España et al., 2005). Schwertmannite, and to a lesser extent jarosite, are the dominant phases controlling Fe(III) solubility in acidic mine drainage, while Al^{3+} solubility is mainly regulated by amorphous to poorly crystalline aluminium phases like basaluminite. Both biotic and abiotic oxidative processes introduce significant quantities of iron and sulphur into the drainage, which subsequently precipitate as iron oxyhydroxysulphates. Schwertmannite and jarosite are the primary sulphate-bearing secondary iron precipitates in such highly acidic environments.

Jarosite precipitation is favoured under extremely acidic conditions (pH < 2.8) in the presence of large quantities of monovalent cations such as K^+, Na^+, and NH_4^+ (España et al., 2005). In contrast, schwertmannite does not require mono- or divalent cations and is commonly not observed in the "hot zones" of AMD generation but in water bodies where highly mineralized AMD waters become diluted to some extent, such as drains, streams, or lakes. Schwertmannite typically precipitates within the pH range of 2.8–4.5, often accompanied by traces of goethite and jarosite as secondary phases. For instance, Dakos et al. (2012) reported that schwertmannite predominated in mine drainage where the pH was buffered around ~3.1–3.7, remaining stable even during heavy rainfall (Figure 1.2).

Similarly, abundant schwertmannite precipitation has been observed at the Tinto Santa Rosa mine (Spain). Effluent flows along narrow channels and stream beds are covered by yellowish and reddish ochreous precipitates, identified predominantly as schwertmannite with minor amounts of goethite and jarosite (Asta et al., 2007). Schwertmannite precipitation is most favourable near the adit mouth, where the pH ranges from 3.0 to 3.4.

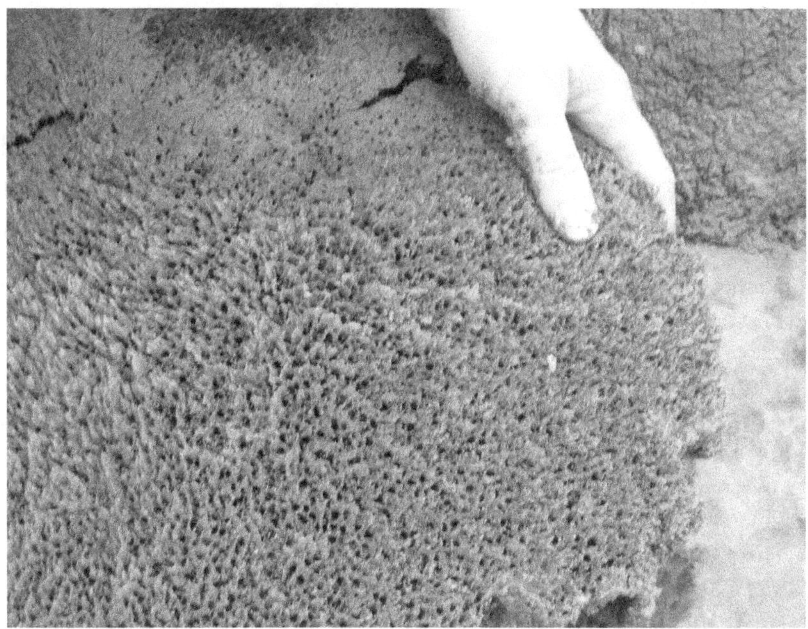

FIGURE 1.2 Images of ochreous precipitates found in the Smolník deposit, East Slovakia (ref. Dakos et al., 2012).

Further downstream, schwertmannite may transform into jarosite and goethite as the pH gradually drops from ~3.5 over just a few metres. Goethite is more commonly found in older Fe(III) deposits at mine sites, as compared to the freshly precipitated colloidal particles.

The pH of AMD environments in which schwertmannite is abundant ranges between 2.5 and 4.0 (Bigham et al., 1996b, Regenspurg et al., 2004) and reflects equilibrium between concentrations of dissolved Fe^{3+} and SO_4^{2-} and schwertmannite being controlled by its solubility (Eq. 1.6, Bigham et al., 1996b).

$$Fe_8O_8(OH)_x(SO_4)_y + (24-2y)H^+ \rightleftharpoons 8Fe^{3+} + ySO_4^{2-} + (24-2y+x)/2H_2O \qquad \text{(Eq. 1.6)}$$

where $8 - x = 2y$ and $1.0 \leq y \leq 1.75$ with the corresponding ion activity product for this dissolution reaction to be (Eq. 1.7).

$$\log IAP = 8 \log a_{Fe^{3+}} + y \log a_{SO_4{}^{2-}} + (24-2y)pH \qquad \text{(Eq. 1.7)}$$

with a_x being the activity of the species x and a range for the ion activity product of (Eq. 1.8)

$$\log IAP = 18.0 \pm 2.5 \qquad \text{(Eq. 1.8)}$$

At higher pH values, schwertmannite tends to become oversaturated. Plots of $\log a_{Fe^{3+}}$ vs. pH for mine waters further show that jarosite forms in the pH range of 1.5–2.5 (Figure 1.3), while ferrihydrite forms at pH levels above 5.5, with schwertmannite occurring in between these ranges (Bigham et al., 1996b). Assuming the $\log K_{sp}$ for goethite is 1.4, jarosite, ferrihydrite, and schwertmannite are metastable compared to goethite at pH values above ~2.0. However, due to the dynamic temporal, spatial, and kinetic factors in mine drainage environments, thermodynamic equilibrium with respect to these secondary minerals is rarely achieved. As a result, mineralogical changes often occur in a sequence of transformation steps (cf. Chapter 2), which could be demonstrated in an alpine stream originating from pyritic schist (Bigham et al., 1996a).

Initially, the stream water was clear but flowed through sediments laden with jarosite. As the stream mixed with a non-acidic tributary 100 m downstream, schwertmannite with traces of goethite began to precipitate, and the pH rose to ~3.0. The stream's appearance changed from clear to ochreous. Eventually, as the stream was fed by more non-acidic tributaries, the pH reached 7.0, and goethite became the dominant mineral phase.

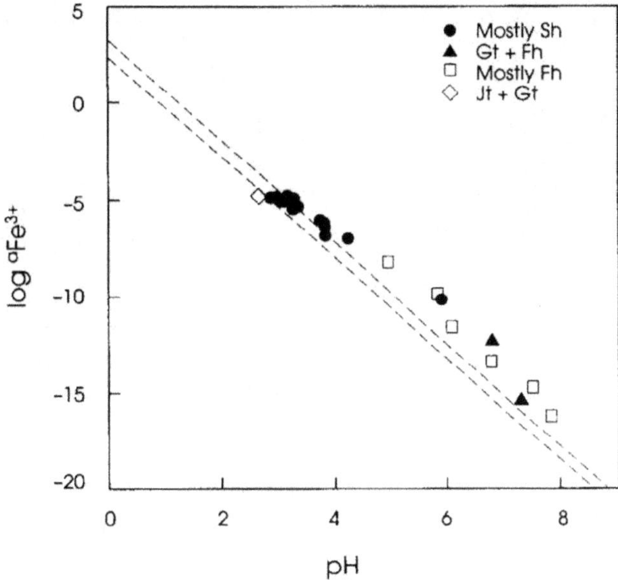

FIGURE 1.3 Plot of log $a_{Fe^{3+}}$ against pH for mine drainage at Ohio, USA (ref. Bigham et al., 1996b). For an average log $a_{SO_4^{2-}}$ of 2.32 found in these waters, a solubility window for schwertmannite can be constrained within log $a_{Fe^{3+}}$ = 3.03–2.60 pH and log $a_{Fe^{3+}}$ = 2.40–2.60 pH [logIAP = 18.0 ± 2.51].

Thermodynamic modelling supports this dynamic behaviour of ferric precipitates. At the source (pH 2.3–2.8), jarosite was in saturation but undersaturated with respect to ferrihydrite and schwertmannite. As the pH rose to 4.0, schwertmannite became the dominant phase, and jarosite became undersaturated. When the pH exceeded 6.0, ferrihydrite reached saturation, replacing schwertmannite as the primary precipitate.

To investigate the reaction kinetics of secondary precipitate formation from AMD, Jönsson et al. (2006) used AMD water from the Kristineberg Zn–Cu mine in Sweden and neutralized it with NaOH to pH 7 and 5.5. The data revealed that OH^- consumption was rapid during the first 20 minutes, followed by a slower phase until about 60–80 minutes, after which it ceased at pH 7 (Figure 1.4). No significant difference in OH^- consumption was observed between temperatures of 10°C and 25°C. The rapid initial consumption of OH^- was attributed to precipitation reactions, while the slower consumption phase was linked to structural rearrangements within the formed precipitates and ongoing precipitation. These structural changes were thought to involve the desorption of SO_4^{2-} from

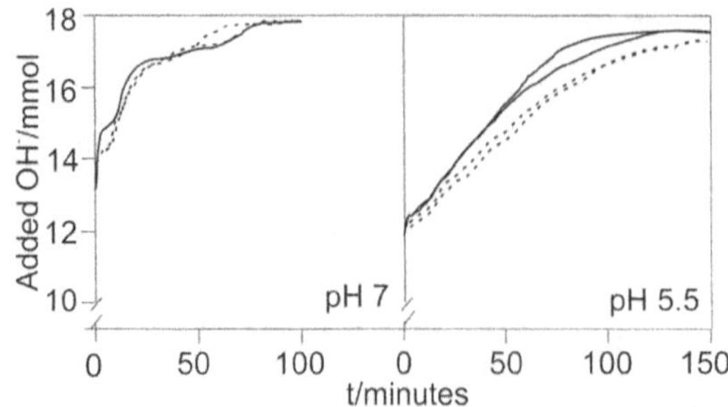

FIGURE 1.4 Hydroxyl ion (OH^-) consumption with respect to time by AMD water from the, Kristineberg Zn–Cu mine, Sweden, at pH 7 and pH 5.5 at 10°C (dash lines) and 25°C (solid lines) (ref. Jönsson et al., 2006).

the surface and structure of the precipitates, an acid-producing process, as well as Fe(II) desorption. Compared to pH 7, slower OH^- consumption was observed at pH 5.5, with lower consumption rates at 10°C than at 25°C. At 25°C, OH^- consumption ceased after 100–120 minutes, while at 10°C, it continued for more than 150 minutes, indicating slower reaction kinetics at temperatures below room temperature (RT) and pH < 7.

At pH 7, lepidocrocite was formed, which increased in crystallinity after 17 days of aging. At pH 5.5, ferrihydrite was initially formed, which transformed into schwertmannite after 17 days, with traces of lepidocrocite and goethite. The surface area of the precipitates decreased from an initial 207 to 105 $m^2\ g^{-1}$ after 17 days of aging at pH 7, likely due to an increase in crystallite size. In contrast, at pH 5.5, the surface area increased slightly from 116 to 136 $m^2\ g^{-1}$, possibly due to slower oxidation, which initially led to slightly larger crystallite sizes.

Kumpulainen et al. (2007) conducted an extensive study on the mineralogy of secondary precipitates and the associated water chemistry of ochreous deposits at seven closed sulphide mines in Finland. They analysed 86 precipitate samples from ditches and ponds surrounding the tailings and waste rock piles. Schwertmannite was the predominant mineral in all the samples, with minor occurrences of goethite in a few cases. Interestingly, the appearance of goethite was linked to the summer months, particularly in July and August, as well as in some cases in November. In contrast, schwertmannite was the sole mineral present at the same sites in May, immediately after the snowmelt in spring. The findings suggest that the schwertmannite formed earlier in the year may slowly transform into goethite during the hotter summer months. This increased goethite content during the summer was further supported by a slight decrease in oxalate-soluble iron compared to total iron, which typically occurs as the crystallinity of a solid phase increases.

1.4.2 Mine pit lakes

Mine pit lakes often form after mining activities cease, leaving behind large, open pits that fill with water. These lakes are typically highly acidic and contain elevated levels of metal(loid)s due to acid generation from exposed sulphide minerals. Some well-studied examples include:

1. Lignite Mine Lake 77, located in the Lusatian mining area of east-central Germany, has been studied extensively (Peine et al, 2000; Ciobotă et al., 2013; Peiffer et al., 2013; Miot et al., 2016). Lake 77 consists of two distinct basins: the central basin, which has a dimictic water regime with seasonal water circulation in spring and autumn, and the smaller northern basin, which is meromictic with permanently anoxic deep waters that do not experience circulation.
2. Mine Pit Lakes in the Iberian Pyrite Belt (IPB), Spain, vary in size and shape, with some spanning up to 1500 m in length and 1000 m in width (Figure 1.5; España et al., 2008; Santofimia et al., 2015). These lakes often show similar stratification patterns with spring-summer stratification and winter turnover, particularly in shallower lakes. Deeper pit lakes, however, may experience permanent stratification with sharp transitions between oxygenated upper layers (mixolimnion) and anoxic bottom layers (monimolimnion).

In these systems, the shallow lakes are often characterized by a metalimnion, a middle layer separating the well-mixed, oxygen-rich upper epilimnion and the isolated, anoxic hypolimnion. Below the metalimnion, dissolved oxygen (DO) and Eh sharply decrease, while Fe(II) concentrations and electrical conductivity (EC) increase, indicating reducing conditions. The bottom monimolimnion remains rich in Fe(II) throughout the year, while the mixolimnion experiences oxygenation and Fe(III) precipitation.

In contrast, holomictic lakes show a complete mixing regime, with an oxygen minimum zone along the metalimnion, where both Fe(II) and total Fe (Fe(T)) are depleted. Dissolved concentrations of metals such as Al, Ca, Zn, Mn, and Co decline in these lower layers, often accompanied by a slight increase in pH.

The pit lakes in the IPB range from circum-neutral (pH 7.2) with low dissolved metal concentrations to highly acidic (pH ~ 2.2) with excess dissolved metal(loid)s (España et al., 2008). The extreme variations in

FIGURE 1.5 Images of mine pit lakes of the Iberian Pyrite Belt (ref. España et al., 2008): Aznalcóllar (top) and San Telmo (bottom).

physicochemical conditions reflect the dynamic interactions between the hydrology, geochemistry, and biology within these lakes.

Schwertmannite is a dominant mineral phase in the sediments of acidic Lake 77 (pH ~ 3.0, Peine et al., 2000). Schwertmannite is formed in the water column (Ciobotă et al., 2013) upon microbial oxidation of Fe^{2+}, which is entering the lake with anoxic groundwater being rich in Fe(II) and SO_4 (Oldham et al., 2019). Acidification of the lakes is following oxidation of Fe^{2+} and subsequent hydrolysis of Fe^{3+} to schwertmannite. In the water body, ferrihydrite forms as a transformation product of freshly formed schwertmannite particles when the pH approaches 5.5, though no goethite has been detected (Miot et al., 2016). In contrast, transformation of schwertmannite to goethite could be clearly observed in the sediment following a pH increase in the pore water from ~3 to ~6 (Peine et al., 2000). Ferrihydrite often contains impurities like Al^{3+} and Si^{4+}, which are incorporated into its structure due to the acidic nature of the lake water, where these ions exist in dissolved form. The formation of ferrihydrite from schwertmannite is attributed to the higher pH (~5.5) and the presence of dissolved Fe^{2+}, which may have triggered the catalytic transformation process. The absence of goethite, despite being the most stable iron mineral, may be due to the association of Al, Si, and OM with ferrihydrite and schwertmannite, as Al^{3+} can substitute for Fe^{3+} within both structures (Miot et al., 2016).

Secondary precipitates of Fe, Al, and Ca under these acidic conditions are likely to form as colloids rather than solid phases, as authigenic precipitation at such low pH levels (pH ~ 2.5–3.0) is sometimes unfavourable (España et al., 2008). Schwertmannite acts as a pH buffer in acidic pit lakes, maintaining a pH range of 2.2–3.5 (Regenspurg et al, 2004; España et al., 2008; Amos et al., 2015; Santofimia et al., 2015). A redox equilibrium is established between Fe(II) and schwertmannite in acidic mine lakes, which implies that schwertmannite is the first mineral precipitates upon Fe(II) oxidation (Regenspurg et al., 2004) and is metastable with goethite and in highly acidic and presence of monovalent cations with jarosite (Figure 1.6).

Schwertmannite precipitation releases H^+, while its dissolution consumes them. Other buffering systems include Al–hydroxysulphates (pH 4.5–5.0) and carbonate systems (CO_3^{2-}–HCO_3^-–$CaCO_3$, pH 7.2) (Amos et al., 2015; Nordstrom et al., 2015; Carrero et al., 2017). Aluminium tends to precipitate when surface water is diluted or neutralized to a pH above 4.5, based on the pKa of Al^{3+}–hydroxides.

In addition to Fe and Al oxyhydroxides, aluminosilicate minerals such as chlorite, biotite, muscovite, kaolinite, plagioclase, and amphiboles also dissolve, releasing significant quantities of Na^+, K^+, Mg^{2+}, Ca^{2+}, and H_4SiO_4. However, unlike Fe and Al oxyhydroxides and carbonates, this dissolution process does not significantly contribute to pH buffering (Amos et al., 2015). These buffering systems influence the mobility of trace metals. Schwertmannite precipitation, acting as an acidic buffer, controls the geochemistry of Fe, As, S, Cu, Pb, and Cr in mine pit lakes, where most contaminants remain dissolved.

Metal concentrations are higher in the bottom layer of pit lakes than in the upper layer due to schwertmannite precipitation in the latter.

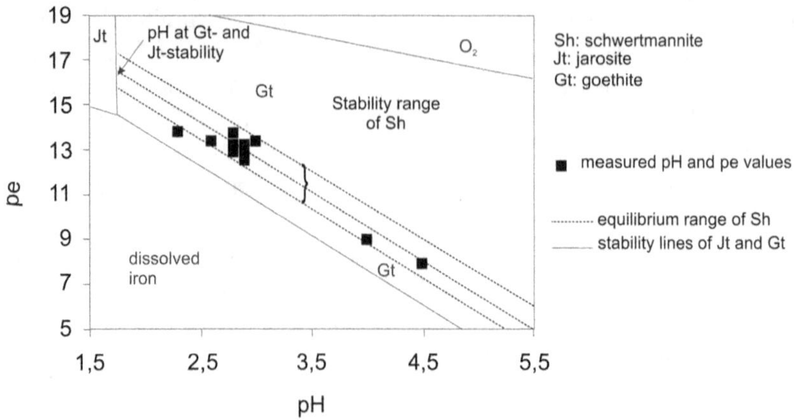

FIGURE 1.6 pE–pH diagram for the K–Fe–SO_4 system. Data points represent measured values of pE and pH from the acidic lignite mining lake ML 77, Germany (ref. Regenspurg et al., 2004).

Schwertmannite formation is typically confined to the upper, oxygenated layers of stratified lakes, while the deeper, anoxic layers limit the availability of Fe^{3+}, necessary for schwertmannite precipitation. In these deeper layers, Fe^{2+} remains the dominant species (Santofimia et al., 2015).

The stratification of pit lakes is not directly related to temperature variations between winter and other seasons. During winter, when surface water temperatures drop, overturning occurs, allowing oxygenation of deeper layers. However, temperature alone does not dictate stratification. In some cases, such as the Nuestra Señora del Carmen (NSC) mine pit lake in Spain, inverted thermal stratification has been observed. In February, for example, surface water was cooler (12.4°C) than the bottom layer (14.3°C), resulting in a significant variation in EC (1.13 vs. 9 mS cm^{-1}).

While Fe(T) concentrations remain relatively constant throughout the year in the bottom layer, Fe(II) and Fe(III) fluctuate due to alternating oxidation and reduction processes. Winter mixing introduces oxygen, allowing Fe(II) to be oxidized to Fe(III) both abiotically and bacterially. In contrast, during the summer, stratification halts oxygen circulation, promoting Fe(III) reduction by iron-reducing bacteria. Since mixing and Fe(II) oxidation occur only briefly during winter, the bottom layer remains Fe(II) rich for most of the year.

1.4.3 Acid sulphate coastal lowlands

Acidity in coastal lowlands is generated by the same mechanism that operates in AMD areas, i.e., the oxidation of sulphide minerals such as pyrite,

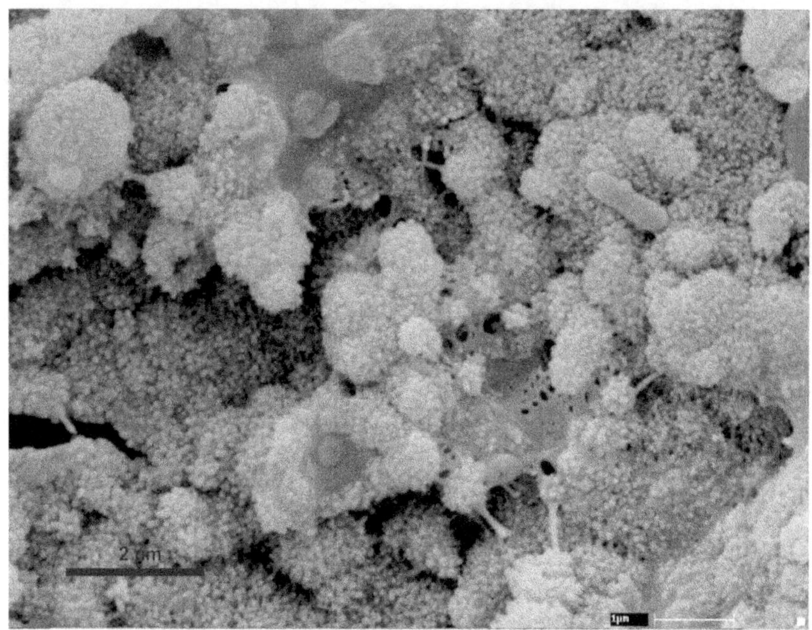

FIGURE 1.7 SEM images showing schwertmannite from acid sulphate soils of eastern Australia with "pin-cushion" morphology (ref. Burton et al., 2006).

pyrrhotite, and marcasite. In addition, sedimentary acid–volatile sulphides may contribute to the acidity and mineral formation (e.g., Sullivan and Bush, 2004).

These sulfidic soils cover approximately 20 million hectares of low-lying coastal areas globally (Burton et al., 2006). Due to the excess sulphur and resulting acidity, these soils are generally unsuitable for agriculture. Schwertmannite is a predominant mineral in these sulfidic lowlands because of the high sulphur and iron content required for its formation, as well as the acidic pH conditions. Burton et al. (2006) observed schwertmannite precipitation with distinct "pin-cushion" morphology in acid sulphate soils from eastern Australia (Figure 1.7).

1.4.4 Polar regions

Schwertmannite occurrence was also reported for polar regions. Following exposure of disseminated pyrite from King George Island, South Shetlands, Antarctica, to oxidation at the interface between the glacier and the sea, schwertmannite has formed resulting in acidic conditions with a pH

FIGURE 1.8 Schwertmannite formed on snow cover by hydrolysis of ferric iron along Cardozo Cove, Antarctica (ref. Dold et al., 2013).

ranging from 3.2 to 4.5 (Dold et al., 2013). This acidity leads to elevated concentrations of Fe, Al, Mn, Zn, and Cu being discharged into the sea.

Schwertmannite, with its characteristic sea urchin-like morphology (Figure 1.8), has also been reported from Cardozo Cove on King George Island. The site had a pH of 3.8 and a dissolved iron concentration of 58.5 μM (Dold et al., 2013). The climatic variability in Antarctica, particularly the extent of snow cover, has been shown to influence the concentration of dissolved iron. During seasons with less snow cover, more mineralized zones are exposed, leading to higher iron concentrations. This differential iron availability directly impacts the population of phytoplankton, as iron is a limiting nutrient in these regions.

Subglacial biochemical and geochemical processes can be inferred from nanoparticles preserved within sediments. Raiswell et al. (2009) reported typical As-rich schwertmannite from Antarctica, with well-preserved "pin-cushion" morphology and characteristic diffraction peaks at 0.255 and 0.151 nm. The schwertmannite was found along with ferrihydrite and goethite. These nanoparticles likely formed under high degrees of supersaturation caused by repeated freezing and thawing cycles through the chemical weathering of pyrite in subglacial sediments being indicative for subglacial water, as water is essential for their development. The edges of the Antarctic continent contain extensive sulphide minerals, which are prone to chemical oxidation by oxygen (Hawkes, 1982). Pyrite oxidation serves as a source of SO_4^{2-} and Fe^{2+}, evidenced by schwertmannite formation. Unlike tropical regions, the oxidation of pyrite is slower in low-temperature conditions like Antarctica. Yet, temperature may remain above freezing year-round beneath the snow cover due to the heat produced by sulphide oxidation, creating microenvironments of water (Dold et al., 2013). High water temperatures are correlated with dissolved Fe^{2+}, and microbial Fe^{2+} oxidation of pyrite may occur during the winter, albeit less efficiently than in the summer. However, schwertmannite (and ferrihydrite) is not ubiquitous across Antarctica, forming only in regions favourable for Fe^{2+} oxidation and Fe^{3+} hydrolysis in the presence of sufficient SO_4^{2-}. Dold et al. (2013) proposed three reasons for the absence of Fe(III)–oxyhydroxides including schwertmannite in some regions: (1) all Fe^{2+} may precipitate in beach sediment as Fe(III)–oxyhydroxides, (2) Fe(II) nanoparticles may migrate through sediment into seawater, and (3) low temperatures may slow Fe(II) oxidation, hindering schwertmannite formation.

Global warming and the resulting glacial melt in the polar regions expose sulphide minerals to atmospheric oxygen, leading to acid rock drainage (ARD) (Dold et al., 2013). Schwertmannite formation in these regions may contribute to iron fluxes that stimulate biological productivity in the surrounding oceans (Raiswell et al., 2009). Calculations by Dold et al. (2013)

estimate an iron input of 3.7×10^9 g Fe year^{-1} from rock weathering along the Antarctic coastline, a figure expected to rise due to ARD. The formation and delivery of schwertmannite and other iron-bearing minerals to the ocean can promote phytoplankton production, enhancing the oxidation of pyrite and increasing iron release. Schwertmannite nanoparticles have also been found in Arctic ice sheets (Raiswell et al., 2009).

1.4.5 Planet Mars

The processes of mineral formation due to weathering of sulphide-bearing ore deposits on Earth may have parallels on Mars. Martian basaltic aquifer fluids, rich in Fe^{2+} from basalt, could have reached the surface through upwelling (Hurowitz et al., 2010). Fe^{2+} would oxidize rapidly, either through photo-oxidation (Eq. 1.9) in the presence of ultraviolet light or by dissolved molecular O_2 (Eq. 1.10), derived from atmospheric H_2O (Eqs. 1.11–1.12), leading to schwertmannite formation (Eq. 1.13). Studies have confirmed the presence of schwertmannite on Mars, potentially formed by acidic groundwater or saline meltwater interacting with the Martian atmosphere (Burns, 1994; Bishop, 1998; Bell et al., 2000; Bibring et al., 2007; Chevrier and Mathé, 2007). Martian soils display strong spectral signatures of schwertmannite, along with ferrihydrite, maghemite, and magnetite (Bishop et al., 2004). Reflectance patterns from Martian soil near Mermaid and Cradle are similar to synthetic schwertmannite, with maximum reflectance at 720 nm and an angular shoulder at 600 nm (Bell et al., 2000). Schwertmannite's visible-region spectra, which resemble those of jarosite and akaganeite, are dominated by a broad Fe^{3+} transition at ~0.91 μm and a local maximum at ~0.73 μm. Its near-infrared (NIR) spectra exhibit brighter features at ~1.45, ~1.95, and ~3 μm, while its mid-infrared (MIR) spectra are darker, with water features at ~6.1 μm and weak SO_4^{2-} and OH^- bands at longer wavelengths (Bishop and Murad, 1996).

$$Fe_{aq}{}^{2+} + H_2O + h\nu = Fe^{3+} + OH^- + 0.5H_2 \quad \text{(Eq. 1.9)}$$

$$Fe_{aq}{}^{2+} + 0.25O_2(aq) + H^+ = Fe_{aq}{}^{3+} + 0.5H_2O \quad \text{(Eq. 1.10)}$$

$$0.5H_2O + h\nu = 0.25O_2 + H^+ + e^- \quad \text{(Eq. 1.11)}$$

$$H_2O + e^- = OH^- + 0.5H_2 \quad \text{(Eq. 1.12)}$$

$$Fe_{aq}{}^{2+} + 0.125SO_4{}^{2-} + 1.75H_2O = FeO(OH)_{0.75}(SO_4)_{0.125} + 2.75H^+ \quad \text{(Eq. 1.13)}$$

Along with schwertmannite, both hematite and goethite have been reported on Mars. These minerals are thought to form through the transformation of schwertmannite into jarosite (Eq. 1.14) and goethite (Eq. 1.15), which can eventually dehydroxylate into hematite (Eq. 1.16). The oxidation of iron and subsequent precipitation of iron minerals is believed to have been sufficient to generate acidity on the Martian surface.

$$\begin{aligned} &FeO(OH)_{0.75}(SO_4)_{0.125} + 0.25H_2O + 0.75H^+ \\ &+0.5417SO_4{}^{2-} + 0.35\,Na^+/K^+ = 0.33(Na, K)Fe_3(SO_4)_2(OH)_6 \end{aligned} \quad \text{(Eq. 1.14)}$$

$$\begin{aligned} &FeO(OH)_{0.75}(SO_4)_{0.125} + 0.25H_2O = \\ &\qquad FeO(OH) + 0.25H^+ + 0.125SO_4{}^{2-} \end{aligned} \quad \text{(Eq. 1.15)}$$

$$FeO(OH) = 0.5Fe_2O_3 + 0.5H_2O \quad \text{(Eq. 1.16)}$$

This finding supports the existence of two separate and chemically distinct aqueous environments on Mars. The first is a subsurface, anoxic hydrosphere buffered to a circum-neutral pH but isolated from the atmosphere. The second is a surface hydrosphere, where rapid oxidation of Fe^{2+} in the presence of atmospheric oxidants drives the system towards acidity. This surface acidity is believed to be associated with the aridification of the Martian environment, which likely occurred over extended periods of time as the planet's climate became drier.

Precipitated schwertmannite has been shown to effectively trap contaminants such as Ni, Zn, and Cr (Zhao and McLennan, 2013). This study highlights improved removal of these metals when schwertmannite forms during iron oxidation processes, eventually leading to the formation of hematite as the final product. A similar phase transformation process occurs on Earth, particularly in mine drainage areas like the Rio Tinto basin in Spain (Fernández-Remolar et al., 2005). The continuous diagenesis of sulphate-rich phases is believed to be a key factor in the development of these minerals.

1.4.6 Natural acidic lakes

Schwertmannite has been identified in several naturally acidic lakes, with Lake Matsuo-Goshikinuma in Japan being a notable example. This lake has a small surface catchment, and its primary water input comes from anoxic groundwater entering through a vent at the lake floor, a result of the dissolution of disseminated pyrite deposits associated with Quaternary volcanism (Childs et al., 1998). The lake's pH remains consistently between 3.0 and 3.2 throughout the year, with no significant vertical variation. The sulphur and iron concentrations fluctuate between 120–130 mg L^{-1} and 20–40 mg L^{-1}, respectively.

The lake's geochemistry is controlled by three major mixing phases annually. Between September and June, the lake undergoes complete and continuous mixing, during which DO is evenly distributed. This period of mixing leads to the oxidation of iron and sulphur, resulting in the lake's turbid brown colour and high bacterial counts, particularly of the iron-oxidizing bacteria *T. ferrooxidans*. By July, the lake's circulation slows, reducing the DO content and limiting the oxidation of Fe^{2+}. The water changes colour, transitioning from greyish-green to bluish-green as oxygen levels decrease. By August, the lake appears clear blue, with visibility extending to ~10 m, and oxidation processes nearly halt due to a dramatic reduction in DO (~10% saturation), except for a thin surface layer approximately 0.5 m deep.

1.4.7 Mine water treatment plants

Acidic mine water treatment plants aim to neutralize highly acidic mine effluents, which are often enriched with toxic elements and sulphate. Neutralization using an alkaline agent is the standard practice in these plants, facilitating the precipitation of secondary minerals. Before the neutralization process, however, acidic mine water can produce schwertmannite through aeration and the oxidation of Fe^{2+} by suitable media.

Childs et al. (1998) reported the presence of schwertmannite in several treatment plants after the aeration of mine water and microbial oxidation by *T. ferrooxidans*, but prior to lime neutralization. For instance, schwertmannite precipitates were obtained from the Matsuo Mine drainage treatment plant, which processes drainage from a pyrite mine, and from the Dowa Mining Corporation, which treats effluents from the Yanahara Mine in western Honshu, Japan.

A pilot plat is set up at lignite open pit mine, Nochten, Germany, to produce schwertmannite upon microbial oxidation of ferrous iron in acidic mine waters at a pH of 2.85–3.1. A maximum load of 2.5 $m^3\ h^{-1}$ and an oxidation rate of 55 g Fe(II) $m^{-3}\ h^{-1}$ were achieved during this treatment process (Janneck et al., 2010).

1.5 Natural attenuation

Leaching of mine wastes not only generates acidity, but also releases large quantities of Fe, Al, Si, SO_4^{2-}, and trace metals. These ions remain dissolved under such acidic conditions. However, as the acidic water flows downslope, it gains alkalinity, leading to the precipitation of oxides and oxyhydroxides of Al and Fe due to a loss of solubility. The ore minerals are processed for economically valuable metals such as Pb, Cu, Zn, Mn, and Fe, but the remaining gangue and waste materials, discarded due to high processing costs, often lead to the oxidation of sulphide minerals within the tailings. Sulphide minerals are thermodynamically unstable when exposed to oxygen and water, leading to their oxidative weathering, which generates H^+ and releases sulphate and metal(loid)s into tailings pore water.

In acidic conditions, metals such as Fe, Al, Zn, and Cu exhibit high dissolved concentrations and greater mobility, while (hydr)oxyanion-forming elements (e.g., As, Se, Mo, and Sb) are often mobile in circum-neutral to alkaline waters. Acidic pore water and drainage may contain high concentrations of (hydr)oxyanion-forming elements (Nordstrom et al., 2000; Majzlan et al., 2014). As a result, AMD commonly exhibits high concentrations of Fe, Al, Cu, Ni, Zn, and Mn, while NMD may contain elevated levels of Ni, Zn, Mn, As, Mo, and Sb (España et al., 2005; Lindsay et al., 2009).

Trace metal transport and attenuation are controlled by precipitation–dissolution, sorption–desorption, and redox reactions. The high dissolved concentrations of Fe and SO_4^{2-} from sulphide oxidation often facilitate the formation of hardpan layers made of Fe (oxy)hydroxides or (hydroxy) sulphates. Gypsum ($CaSO_4{\cdot}2H_2O$) precipitation often plays a significant role in controlling dissolved SO_4^{2-} concentrations in carbonate-rich tailings. Redox reactions strongly influence the mobility of redox-sensitive elements like S, Fe, As, and Se, which have multiple oxidation states. Sulphate reduction can promote the precipitation of metal-sulphide phases, while the reduction of As(V) to As(III) typically enhances As mobility. Under moderately acidic to circum-neutral conditions, the mobility of dissolved metal(loid) oxyanions, including $H_2AsO_4^-$, SeO_4^{2-}, and $Sb(OH)_6^-$, is limited by sorption onto positively charged mineral surfaces.

For example, at the Paroistenjärvi Mine in Ylőjärvi, Finland, sulphide minerals such as pyrite, chalcopyrite, arsenopyrite, and pyrrhotite have been documented, along with a variety of ferric oxyhydroxides and oxyhydroxysulphates. Schwertmannite is predominant along the mine ditches at pH ~ 3.0, with ferrihydrite and goethite present where the drainage pH exceeds 5. The drainage water contains elevated levels of As, Mn, Cu, and Zn, and the secondary precipitates contain significant amounts of these elements (Carlson et al., 2002). Similar results have been reported from the Smolník abandoned mine in East Slovakia, with high concentrations of As, Cu, Mn, Zn, Pb, Cr, Sb, and Ni in schwertmannite (Dakos et al., 2012).

Interestingly, Asta et al. (2007) found high concentrations of As, Co, Ni, and Cd near the adit mouth of the Tinto Santa Rosa mine in Spain's

IPB, with a systematic decrease in As concentrations downstream. Other metals, such as Cd, Ni, Co, and Pb, did not exhibit a similar trend, possibly due to their positive charge at the drainage pH, while As exists as a negatively charged species ($H_2AsO_4^-$). Jarosite has also shown excellent uptake of As, V, and Sr at the Libiola Fe–Cu mine in Italy, suggesting it can act as a potential natural sink for contaminants (Marescotti et al., 2012). Similarly, ferrihydrite and goethite are enriched with Zn, V, Ni, and Cu, while Al-enriched amorphous phases contain the highest concentrations of trace metals at this mine site.

1.6 Summary

Oxidative dissolution of sulphide minerals such as FeS_2, FeAsS, and PbS along mine waste sites generates Fe^{2+}, which is further oxidized to Fe^{3+}. The generated Fe^{3+} enhances sulphide oxidation kinetics five to six times greater than that occurs by atmospheric O_2. Presence of iron-oxidizing bacteria such as *A. ferrooxidans* acts as catalysts for Fe^{3+} generation that ultimately triggers sulphide oxidation until this oxidative dissolution process achieves saturation. The ferric oxyhydroxides and/or oxyhydroxysulphates precipitated upon hydrolysis of Fe^{3+} are the main pathway of schwertmannite formation around mine drainage environments. Because of its amorphous nature, it was being identified as amorphous ferric hydroxides before its first recognition in 1990s. Schwertmannite is the most common secondary ferric precipitate around AMD localities within a pH range of 2.5–4.5 and SO_4^{2-} concentrations of 1000–3000 mg L^{-1}. Most common secondary precipitates found together with schwertmannite are jarosite and goethite either by direct precipitation from mine drainage or through transformation of schwertmannite.

Besides AMD, schwertmannite also occurs along acid sulphate coastal lowlands as widely reported from eastern Australia through oxidation of FeS_2, FeS, etc. Disseminated pyrite from King George Island, South Shetlands, Antarctica, has been documented and presence of abundant pyrite in many places of Antarctica. Global warming and the resulting glacial melt in the polar regions expose sulphide minerals to atmospheric oxygen, leading to ARD. These sulphide minerals act as sources for Fe^{3+} and SO_4^{2-} for schwertmannite formation. Schwertmannite has also been reported from Arctic regions. Martian soils display strong spectral signatures of schwertmannite, along with ferrihydrite, maghemite, and magnetite. Martian basaltic aquifer fluids that are rich in Fe^{2+} from basalt would oxidize rapidly through photo-oxidation or by dissolved molecular O_2 derived from atmospheric H_2O forming schwertmannite, which is believed as the major mechanism of its formation in Mars. Formation of schwertmannite along mine pit lakes has also been reported from Lake 77, Germany, and IPB, Spain. Oxidation of sulphides along these abandoned open cast mine sites serves as the source of Fe^{3+} and SO_4^{2-}. Natural acidic lakes that are not associated with mine drainage, e.g., Lake Matsuo-Goshikinuma, Japan, also favour precipitation of schwertmannite where Fe^{3+} and SO_4^{2-} are derived from dissolution of FeS_2 associated with quaternary volcanism. Schwertmannite occurrence has also been documented along mine water treatment plants.

The precipitated schwertmannite is often not pure, but associated with trace metals and other impurities like Al^{3+} and Si^{4+} because of its strong sorption efficiency. The occurrence of schwertmannite always acts a natural sink of pollutants through surface adsorption and ion exchange process.

References

Amos RT, Blowes DW, Bailey BL, Sego DC, Smith L and Ritchie AIM (2015). Waste-rock hydrogeology and geochemistry. *Applied Geochemistry* 57: 140–156.

Asta MP, Cama J, Gault AG, Charnock JM and Queralt I (2007). Characterization of AMD sediments in the discharge of the Tinto santa rosa mine (Iberian Pyritic Belt, SW Spain). In: Cidu R and Frau F (Eds), *IMWA symposium on water in mining environments*. IMWA, Cagliari. pp. 391–395.

Baker BJ and Banfield JF (2003). Microbial communities in acid mine drainage. *FEMS Microbiology Ecology* 44: 139–152.

Bell III JF, McSween Jr HY, Crisp JA, Morris RV, Murchie SL, Bridges NT, Johnson JR, Britt DT, Golombek MP, Moore HJ, Ghosh A, Bishop JL, Anderson RC, Brückner J, Economou T, Greenwood JP, Gunnlaugsson HP, Hargraves M, Hviid S, Knudsen JM, Madsen MB, Reid R, Rieder R and Soderblom L (2000). Mineralogical and compositional properties of Martian soil and dust: Results from Mars pathfinder. *Journal of Geophysical Research* 105: 1721–1755.

Bibring JP, Arvidson RE, Gendrin A, Gondet B, Langevin Y, Le Mouelicn S, Mangold N, Morris RV, Mustard JF, Poulet F, Quantin C and Sotin C (2007). Coupled ferric oxides and sulfates on the Martian surface. *Science* 317: 1206–1210.

Bigham JM, Carlson L and Murad E (1994). Schwertmannite, a new iron oxy-hydroxysulfate from Pyhasalmi, Finland, and other localities. *Mineralogical Magazine* 393: 641–648.

Bigham JM and Nordstrom DK (2000). Iron and aluminum hydroxysulfates from acid sulfate waters. In: Alpers CN, Jambor JL and Nordstrom DK (Eds), *Reviews in mineralogy and geochemistry, sulfate minerals: Crystallography, geochemistry, and environmental significance*, vol. 40. The Mineralogical Society of America, pp. 351–403.

Bigham JM, Schwertmann U, Carlson L and Murad E (1990). A poorly crystallized oxyhydroxysulfate of iron formed by bacterial oxidation of Fe(II) in acid mine waters. *Geochimica et Cosmochimica Acta* 54: 2743–2758.

Bigham JM, Schwertmann U and Pfab G (1996a). Influence of pH on mineral speciation in a bioreactor stimulating acid mine drainage. *Applied Geochemistry* 11: 845–849.

Bigham JM, Schwertmann U, Traina SJ, Winland RL and Wolf M (1996b). Schwertmannite and the chemical modeling of iron in acid sulfate waters. *Geochimica et Cosmochimica Acta* 60: 2111–2121.

Bishop JL (1998). Biogenic catalysis of soil formation on mars? *Origins of Life and Evolution of the Biosphere* 28: 449–459.

Bishop JL, Dyar MD, Lane MD and Banfield JF (2004). Spectral identification of hydrated sulfates on Mars and comparison with acidic environments on Earth. *International Journal of Astrobiology* 3: 275–285.

Bishop JL and Murad E (1996). Schwertmannite on Mars? Spectroscopic analyses of schwertmannite, its relationship to other ferric minerals, and its possible presence in the surface material on Mars. In: Mineral spectroscopy: A tribute to Roger G. Burns, edited by Dyar MD, McCammon C and Schaefer MW. *Special Publication of Geochemical Society* 5: 337–358.

Blodau C (2006). A review of acidity generation and consumption in acidic coal mine lakes and their watersheds. *Science of the Total Environment* 369: 307–332.

Bond PL, Druschel GK and Banfield JF (2000). Comparison of acid mine drainage microbial communities in physically and geochemically distinct ecosystems. *Applied and Environmental Microbiology* 66: 4962–4971.

Boukhalfa C and Chaguer M (2012). Characterisation of sediments polluted by acid mine drainage in the northeast of Algeria. *International Journal of Sediment Research* 27: 402–407.

Brady KS, Bigham JM, Jaynes JF and Logan TJ (1986). Influence of sulfate on fe-oxide formation: Comparisons with a stream receiving acid mine drainage. *Clays and Clay Minerals* 3: 266–274.

Burns RG (1994). Schwertmannite on mars: Deposition of this ferric oxyhydroxysulfate mineral in acidic saline meltwaters. In: *Proceedings of lunar and planetary sciences XXV*. Lunar and Planetary Institute, Houston, TX, pp. 203–204.

Burton ED, Bush RT and Sullivan LA (2006). Sedimentary iron geochemistry in acidic waterways associated with coastal lowland acid sulfate soils. *Geochimica et Cosmochimica Acta* 70: 5455–5468.

Burton ED, Bush RT, Sullivan LA and Mitchell DRG (2007). Reductive transformation of iron and sulfur in schwertmannite-rich accumulations associated with acidified coastal lowlands. *Geochimica et Cosmochimica Acta* 71: 4456–4473.

Caraballo MA, Rimstidt JD, Macías F, Nieto JM and Hochella Jr. MF (2013). Metastability, nanocrystallinity and pseudo-solid solution effects on the understanding of schwertmannite solubility. *Chemical Geology* 360–361: 22–31.

Carlson C, Bigham JM, Schwertmann U, Kyek A and Wagner F (2002). Scavenging of As from acid mine drainage by schwertmannite and ferrihydrite: A comparison with synthetic analogues. *Environmental Science &Technology* 36: 1712–1719.

Carrero S, Fernandez-Martinez A, Pérez-López R, Poulain A, Salas-Colera E and Nieto JM (2017). Arsenate and selenate scavenging by basaluminite: Insights into the reactivity of aluminum phases in acid mine drainage. *Environmental Science & Technology* 51: 28–37.

Chevrier V and Mathé PE (2007). Mineralogy and evolution of the surface of Mars: A review. *Planetary and Space Science* 55: 289–314.

Childs CW, Inoue K and Mizota C (1998). Natural and anthropogenic schwertmannites from Towada-Hachimantai National Park, Honshu, Japan. *Chemical Geology* 144: 81–86.

Ciobotă V, Lu S, Tarcea N, Rösch P, Küsel K and Popp J (2013). Quantification of the inorganic phase of the pelagic aggregates from an iron contaminated lake by means of Raman spectroscopy. *Vibrational Spectroscopy* 68: 212–219.

Dakos Z, Kupka D, Kovařík M, Jablonovská K, Krištúfek V and Achimovičová M (2012). Secondary iron minerals present in AMD sediments from Smolník abandoned mine. *Nova Biotechnologica et Chimica* 11: 87–92.

Dinelli E and Tateo F (2002). Different types of fine-grained sediments associated with acid mine drainage in the Libiola Fe–Cu mine area (Ligurian Apennines, Italy). *Applied Geochemistry* 17: 1081–1092.

Dold B (2014). Evolution of acid mine drainage formation in sulphidic mine tailings. *Minerals* 4: 621–641.

Dold B, Gonzalez-Toril E, Aguilera A, Lopez-Pamo E, Cisternas ME, Bucchi F and Amils R (2013). Acid rock drainage and rock weathering in Antarctica: Important sources for iron cycling in the southern ocean. *Environmental Science & Technology* 47: 6129–6136.

Dzombak DA and Morel FMM (1990). *Surface complexation modeling: Hydrous ferric oxide.* John Wiley & Sons, pp. 416.

Edwards KJ, Bond P, Gihring TM and Banfield JF (2000). An archaeal iron-oxidizing extreme acidophile important in acid mine drainage. *Science* 287: 1796–1799.

España JS, López Pamo E, Pastor ES, Diez Ercilla M (2008). The acidic mine pit lakes of the Iberian Pyrite Belt: An approach to their physical limnology and hydrogeochemistry. *Applied Geochemistry* 23: 1260–1287.

España JS, López Pamo E, Santofimia E, Aduvire O, Reyes J and Barettino D (2005). Acid mine drainage in the Iberian Pyrite Belt (Odiel river watershed, Huelva, SW Spain): Geochemistry, mineralogy and environmental implications. *Applied Geochemistry* 20: 1320–1356.

España JS, López Pamo E, Santofimia E, Reyes Andrés J and Martín Rubí JA (2006). The removal of dissolved metals by hydroxysulphate precipitates during oxidation and neutralization of acid mine waters, Iberian Pyrite Belt. *Aquatic Geochemistry* 12: 269–298.

Evangelou VP and Zhang YL (1995). A review: Pyrite oxidation mechanisms and acid mine drainage prevention. *Critical Reviews in Environmental Science & Technology* 25: 141–199.

Fernández-Remolar DC, Morris RV, Gruener JE, Amils R and Knoll AH (2005). The Rıo Tinto Basin, Spain: Mineralogy, sedimentary geobiology, and implications for interpretation of outcrop rocks at Meridiani Planum, Mars. *Earth and Planetary Science Letters* 240: 149–167.

French RA, Caraballo MA, Kim B, Rimstidt JD, Murayama M and Hochella Jr. MF (2012). The enigmatic iron oxyhydroxysulfate nanomineral schwertmannite: Morphology, structure, and composition. *American Mineralogist* 97: 1469–1482.

Hawkes DD (1982). Nature and distribution of metalliferous mineralization in the northern Antarctic Peninsula. *Journal of Geological Society of London* 139: 803–809.

Hurowitz JA, Fischer WW, Tosca NJ and Milliken RE (2010). Origin of acidic surface waters and the evolution of atmospheric chemistry on early Mars. *Nature Geosciences* 3: 323–326.

Jamieson HE, Walker SR, Parsons MB (2015). Mineralogical characterization of mine waste. *Applied Geochemistry* 57: 85–105.

Janneck E, Arnold I, Koch T, Meyer J, Burghardt D and Ehinger S (2010). Microbial synthesis of schwertmannite from lignite mine water and its utilization for removal of arsenic from mine waters and for production of iron pigments. In: *Mine water and innovative thinking*. IMWA, Sydney, pp. 83–86.

Jönsson J, Jönsson J and Lövgren L (2006). Precipitation of secondary Fe(III) minerals from acid mine drainage. *Applied Geochemistry* 21: 437–445.

Jönsson J, Persson P, Sjoberg S and Lovgren L (2005). Schwertmannite precipitated from acid mine drainage: Phase transformation, sulphate release and surface properties. *Applied Geochemistry* 20: 179–191.

Karathanasis AD, Evangelou VP and Thompson YL (1988). Aluminum and iron equilibria in soil solutions and surface waters of acid mine watersheds. *Journal of Environmental Quality* 17: 534–543.

Kim JJ, Kim SJ and Tazaki K (2002). Mineralogical characterization of microbial ferrihydrite and schwertmannite, and non-biogenic Al-sulfate precipitates from acidmine drainage in the Donghae mine area, Korea. *Environmental Geology* 42: 19–31.

Kumpulainen S, Carlson L and Raisanen ML (2007). Seasonal variations of ochreous precipitates in mine effluents in Finland. *Applied Geochemistry* 22: 760–777.

Küsel K (2003). Microbial cycling of iron and sulfur in acidic coal mining lake sediments. *Water, Air, and Soil Pollution* 3: 67–90.

Küsel K and Dorsch T (2000). Effect of supplemental electron donors on the microbial reduction of Fe(III), sulfate, and CO_2 in coal mining–impacted freshwater lake sediments. *Microbial Ecology* 40: 238–249.

Lazaroff N (1963). Sulfate requirement for iron oxidation by *Thiobacillus ferrooxidans*. *Journal of Bacteriology* 85: 78–83.

Lazaroff N, Signal W and Wasserman A (1982). Iron oxidation and precipitation of ferric hydroxysulfates by resting *Thiobacillus ferrooxidans* cells. *Applied and Environmental Microbiology* 43: 924–938.

Lee G, Bigham JM and Faure G (2002). Removal of trace metals by coprecipitation with Fe, Al and Mn from natural waters contaminated with acid mine drainage in the Ducktown Mining District, Tennessee. *Applied Geochemistry* 17: 569–581.

Lindsay MBJ, Condon PD, Jambor JL, Lear KG, Blowes DW and Ptacek CJ (2009). Mineralogical, geochemical, and microbial investigation of a sulfide-rich tailings deposit characterized by neutral drainage. *Applied Geochemistry* 24: 2212–2221.

Lindsay MBJ, Moncur MC, Bain JG, Jambor JL, Ptacek CJ, Blowes DW (2015). Geochemical and mineralogical aspects of sulfide mine tailings. *Applied Geochemistry* 57: 157–177.

Majzlan J, Plášil J, Škoda R, Gescher J, Kögler F, Rusznyak A, Küsel K, Neu TR, Mangold S and Rothe J (2014). Arsenic-rich acid mine water with extreme arsenic concentration: mineralogy, geochemistry, microbiology, and environmental implications. *Environmental Science & Technology* 48: 13685–13693.

Marescotti P, Carbone C, Comodi P, Frondini F and Lucchetti G (2012). Mineralogical and chemical evolution of ochreous precipitates from the Libiola Fe–Cu-sulfide mine (Eastern Liguria, Italy). *Applied Geochemistry* 27: 577–589.

Miot J, Lu S, Morin G, Adra A, Benzerara K and Küsel K (2016). Iron mineralogy across the oxycline of a lignite mine lake. *Chemical Geology* 434: 28–42.

Moncur MC, Ptacek CJ, Blowes DW, Jambor JL (2005). Release, transport and attenuation of metals from an old tailings impoundment. *Applied Geochemistry* 20: 639–659.

Murad E and Rojík P (2003). Iron-rich precipitates in a mine drainage environment: Influence of pH on mineralogy. *American Mineralogist* 88: 1915–1918.

Murad E and Rojík P (2004). Jarosite, schwertmannite, goethite, ferrihydrite and lepidocrocite: The legacy of coal and sulfide ore mining. In: Singh B (Ed), *SuperSoil 2004: 3rd Australian New Zealand soils conference*, *SuperSoil 2004, Sydney, Australia,* pp. 1–8.

Nordstrom DK (1982). Aqueous pyrite oxidation and the consequent formation of secondary iron minerals. In: Kittrick JA, Fanning DS and Hosner LR (Eds), *Acid sulfate weathering*. Soil Science Society of America, pp. 37–56.

Nordstrom DK (2003). Effects of microbiological and geochemical interactions in mine drainage. In: Jambor JL, Blowes DW and Ritchie AIM (Eds), *Environmental aspects of mine wastes*. Short Course Series, vol. 31. Mineralogical Association of Canada, pp. 227–238.

Nordstrom DK (2011). Hydrogeochemical processes governing the origin, transport and fate of major and trace elements from mine wastes and mineralized rock to surface waters. *Applied Geochemistry* 26: 1777–1791.

Nordstrom DK, Alpers CN, Ptacek CJ and Blowes DW (2000). Negative pH and extremely acidic mine waters from iron mountain, California. *Environmental Science & Technology* 34: 254–258.

Nordstrom DK, Blowes DW and Ptacek CJ (2015). Hydrogeochemistry and microbiology of mine drainage: An update. *Applied Geochemistry* 57: 3–16.

Nordstrom DK and Southam G (1997). Geomicrobiology of sulfide mineral oxidation. In: Banfield JF and Nealson KN (Eds), *Geomicrobiology: Interactions between microbes and minerals*. Mineralogical Society of America, pp. 361–390.

Oldham C, Beer J, Blodau C, Fleckenstein J, Jones L, Neumann C and Peiffer S (2019). Controls on iron(II) fluxes into waterways impacted by acid mine drainage: A Damköhler analysis of groundwater seepage and iron kinetics. *Water Research* 153: 11–20.

Peiffer S, Knorr K and Blodau C (2013). The role of iron minerals for the biogeochemistry of acid pit lakes. In: Geller W, Schultze M, Kleinmann R and Wolkersdorfer C (Eds), *Acidic pit lakes: The legacy of coal and metal surface mines*. Springer, pp. 57–62.

Peine A, Küsel K, Tritschler A and Peiffer S (2000). Electron flow in an iron-rich acidic sedimant - Evidence for an acidity-driven iron cycle. *Limnology and Oceanography* 45: 1077–1087.

Raiswell R, Benning LG, Davidson L, Tranter M and Tulaczyk S (2009). Schwertmannite in wet, acid, and oxic microenvironments beneath polar and polythermal glaciers. *Geology* 37: 431–434.

Regenspurg S, Brand A and Peiffer S (2004). Formation and stability of schwertmannite in acidic mining lakes. *Geochimica et Cosmochimica Acta* 68: 1185–1197.

Santofimia E, Löpez-Pamo E and Montero E (2015). Selective precipitation of schwertmannite in a stratified acidic pit lake of Iberian Pyrite Belt. *Mineralogical Magazine* 79: 497–513.

Schippers A, Jozsa PG and Sand W (1996). Sulfur chemistry in bacterial leaching of pyrite. *Applied and Environmental Microbiology* 62: 3424–3431.

Schwertmann U, Bigham JM and Murad E (1995). The first occurrence of schwertmannite in a natural stream environment. *European Journal of Mineralogy* 7: 547–552.

Sidenko NV and Sherriff BL (2005). The attenuation of Ni, Zn and Cu, by secondary Fe phases of different crystallinity from surface and ground water of two sulfide mine tailings in Manitoba, Canada. *Applied Geochemistry* 20: 1180–1194.

Singer PC and Stumm W (1970). Acid mine drainage: The rate determining step. *Science* 167: 1121–1123.

Sullivan LA and Bush RT (2004). Iron precipitate accumulations associated with waterways in drained coastal acid sulfate landscapes of eastern Australia. *Marine and Freshwater Research* 55: 727–736.

Sullivan PJ, Reddy KJ and Yelton JL (1988). Solubility relationships of aluminum and iron minerals associated with acid mine drainage. *Environmental Geological Water Science:* 11: 283–287.

Williams DJ, Bigham JM, Cravotta CA, Traina SJ, Anderson JE and Lyon JG (2002). Assessing mine drainage pH from the color and spectral reflectance of chemical precipitates. *Applied Geochemistry* 17: 1273–1286.

Zhao YS and McLennan SM (2013). Behavior of Ni, Zn and Cr during low temperature aqueous Fe oxidation processes on Mars. *Geochimica et Cosmochimica Acta* 109: 365–383.

2

STRUCTURE AND SOLUBILITY OF SCHWERTMANNITE

2.1 Introduction

Schwertmannite is an amorphous iron sulphate mineral composed primarily of iron (Fe) and sulphur (S), which has posed challenges for its precise identification due to its non-crystalline nature. The difficulties in characterizing schwertmannite stem from its poor crystallinity, small particle size, and amorphous structure, limiting advanced structural studies through techniques like X-ray diffraction (XRD). Despite its amorphous nature, significant progress has been made in understanding schwertmannite through tools like high-resolution transmission electron microscopy (HRTEM) and Mössbauer spectroscopy. These techniques have revealed important characteristics, but the structure of schwertmannite remains somewhat ambiguous and debatable even after three decades of its recognition.

When first described by Brady et al. (1986) as an unidentified ferric sulphate phase in acidic mine drainage environment, schwertmannite was not yet formally named. It was later characterized and named schwertmannite in the 1990s by Bigham et al. (1990, 1994) in honour of Professor Dr. Udo Schwertmann, a geoscientist renowned for his contributions to iron oxide research. Early studies suggested that schwertmannite's structure resembled akaganéite based on diffractograms, infrared (IR), and Mössbauer spectroscopic data, though more precise structural characterization remains elusive due to its poorly ordered nature (Bigham et al., 1990).

Schwertmannite forms under acidic conditions typically within a narrow pH range of 2.8–4.5, and its formation is influenced by the environmental pH. Below this range, jarosite is more likely to form, while higher

DOI: 10.1201/9781003583875-2

pH levels favour the formation of ferrihydrite or goethite. This narrow pH window means that schwertmannite is often a metastable phase and can easily transform into other mineral phases under changing environmental conditions.

2.2 Structure of schwertmannite

2.2.1 Structure akin to akaganéite

It was early proposed that schwertmannite's structure is similar to that of akaganéite, a ferric oxyhydroxide mineral forming in highly acidic (pH < 2.5) Cl^--rich environments (Bigham et al. 1990). Contrary to akaganéite, the characteristic (110) XRD reflection at 0.747 nm was missing in schwertmannite, suggesting structural differences. The row of (O^{2-}, OH^-) anions characteristic for akaganéite is missing along the crystallographic b direction, forming narrow tunnels in the structure. Schwertmannite is believed to have a similar tunnel structure, which is stabilized by the presence of sulphate ions (SO_4^{2-}).

In schwertmannite, SO_4^{2-} shares two oxygen atoms (O^{2-}, OH^-) along the walls of these tunnels, while the other two oxygen atoms are situated along the centre of the tunnels. The role of SO_4^{2-} in schwertmannite's structure is thought to be analogous to that of Cl^- in akaganéite, where SO_4^{2-} resides in the tunnels. The maximum structural uptake of SO_4^{2-} in schwertmannite corresponds to the formula $Fe_8O_8(OH)_6SO_4$, while in akaganéite, the formula is $Fe_8O_8(OH,Cl)_8$. The tunnel in akaganéite has a cross-section of approximately 4 × 4 Å, which is large enough to accommodate Cl^-, OH^-, or water, but too small for SO_4^{2-}. To fit SO_4^{2-} into these structural cavities, the sulphate ions share oxygen atoms with surrounding iron atoms, causing significant distortion in the structure. This distortion likely accounts for the absence of the akaganéite (110) reflection in schwertmannite. The sulphate ion forms a bridged bidentate –Fe–O–SO_2–O–Fe– complex within the tunnels, which may require a doubling of the c parameter (in the tetragonal setting) of akaganéite.

X-ray absorption spectroscopic studies further supported the idea that schwertmannite's structure resembles that of akaganéite but with a large proportion of structural defects (Waychunas et al., 2001). These defects are believed to be proportional to the SO_4^{2-} content, resulting in significant variations in schwertmannite's structural stability and morphology. The incorporation of sulphate into the tunnel structure plays a critical role in shaping schwertmannite's unique characteristics compared to akaganéite.

The tetragonal unit cell of akaganéite has lattice parameters a_O = 1.054 nm and c_O = 0.303 nm, and it is composed of double chains of $FeO_3(OH)_3$

octahedra. These octahedra share corners to create square tunnels with a cross-sectional area of 0.5 × 0.5 nm^2, extending along the 'c' crystallographic axis. Each tunnel contains adjoining cavities formed by eight hydroxyl groups, with one cavity per unit cell. These cavities can accommodate ions or molecules with diameters up to 0.35 nm, and the necks of these cavities have a diameter of approximately 0.27 nm.

In the case of akaganéite, Cl^- ions, with a diameter of around 0.36 nm, are small enough to fit within these cavities, even though they are slightly larger than the cavity neck diameter. In contrast, the SO_4^{2-} ion, which has a diameter of 0.50 nm, is significantly larger and therefore cannot fit into the tunnels of akaganéite without modifying the structure.

Because of this size disparity, it is assumed that SO_4^{2-} in schwertmannite cannot occupy these cavities directly but must share oxygen atoms with the surrounding iron atoms to be accommodated. Specifically, the hydroxyl groups lining the faces of the tunnels, rather than the oxygen atoms comprising the tunnel corners, are responsible for facilitating such bridging between SO_4^{2-} and the iron atoms.

The distances between the centres of the exposed hydroxyl groups in the tunnels are 0.30 nm on the same face, 0.35 nm on adjacent faces, and 0.50 nm on opposite faces (Figure 2.1). These dimensions are compared to the O–O distance in free SO_4^{2-}, which is 0.26 nm. For SO_4^{2-} to bond with Fe atoms in the tunnel, it would cause structural distortions. Bonding

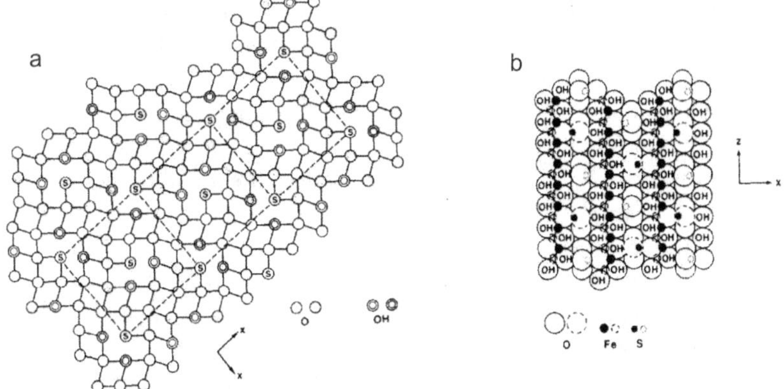

FIGURE 2.1 Proposed structural model of schwertmannite (a) perpendicular to "z" with open and shaded circles indicating oxygens or hydroxyls occurring in and below the plane of the paper, respectively. Circles labelled "s" denote tunnel oxygens comprising part of the SO_4 ion; (b) parallel to "z" with solid and broken lines indicating atoms occurring in and below the plane of the paper, respectively. Both drawings are shown without distortion (ref. Bigham et al., 1990).

between the sulphate ion and iron atoms on adjacent faces of the cavity would allow for the formation of a bridged bidentate complex of the type –Fe–O–SO_2–O–Fe–, where two free oxygen atoms from the sulphate ion could occupy two adjacent cavities along the tunnel axis. This configuration allows the sulphate ion to bond within the structure, but the necessary distortion of the tunnel and cavity structure is consistent with the poor crystallinity of schwertmannite. Therefore, the bonding of SO_4^{2-} in schwertmannite's structure causes significant distortion, explaining why the crystallinity of schwertmannite is poor compared to that of akaganéite. This distortion helps account for the structural differences between these two iron oxyhydroxide minerals.

This bidentate bridging complex is supported by the appearance of S–O IR frequencies at <1200 cm^{-1}. Raman evidence for structural SO_4 was presented by Mazzetti and Thistlethwaite (2002), who attributed the splitting of the $v_3(SO_4^{2-})$ band to bridging bidentate sulphate anions. This interpretation is similar to that of Bigham et al. (1990), who suggested that the IR spectra of schwertmannite were comparable to those of akaganéite, but with superimposed Fe–SO_4 features from structural and adsorbed SO_4 (Figure 2.2). Two distinctly different Fe sites are essential for the formation of the bidentate bridging complex: one where Fe is in normal octahedral coordination with O and OH as $FeO_3(OH)_3$, and the other where OH is replaced by SO_4 to yield $FeO_3(OH)_2O$–SO_3. This is evidenced by Mössbauer spectra, which show two distributions of quadrupole splitting in the paramagnetic state and two distributions of magnetic hyperfine fields with different quadrupole interactions in the magnetically ordered state (Bigham et al., 1990).

Each O atom is coordinated to three Fe atoms in the tunnel structure of akaganéite and is not reactive for ligand exchange reactions such as sulphate adsorption. Since schwertmannite shares a similar tunnel structure to akaganéite, sulphate will not be able to form inner-sphere complexes inside the tunnel unless the double chains of the edge-sharing Fe octahedra have some structural defects compared to akaganéite. Therefore, singly-Fe coordinated functional groups (i.e., $\equiv Fe–OH^{-1/2}$ and $\equiv Fe–OH_2^{+1/2}$) form for ligand exchange with sulphate. The defects in schwertmannite could also contribute to the distortion and enlargement of the tunnels, making them spacious enough to host sulphate ions, which are larger than Cl^-. The conversion of sulphate outer-sphere complexes to inner-sphere complexes, driven by sample dehydration, suggests that wet schwertmannite samples contain water molecules inside the tunnel in addition to sulphate ions. The reversibility of this conversion indicates that the tunnel readily undergoes hydration and dehydration. The tunnel water molecules may form hydrogen bonds with both the schwertmannite structural framework

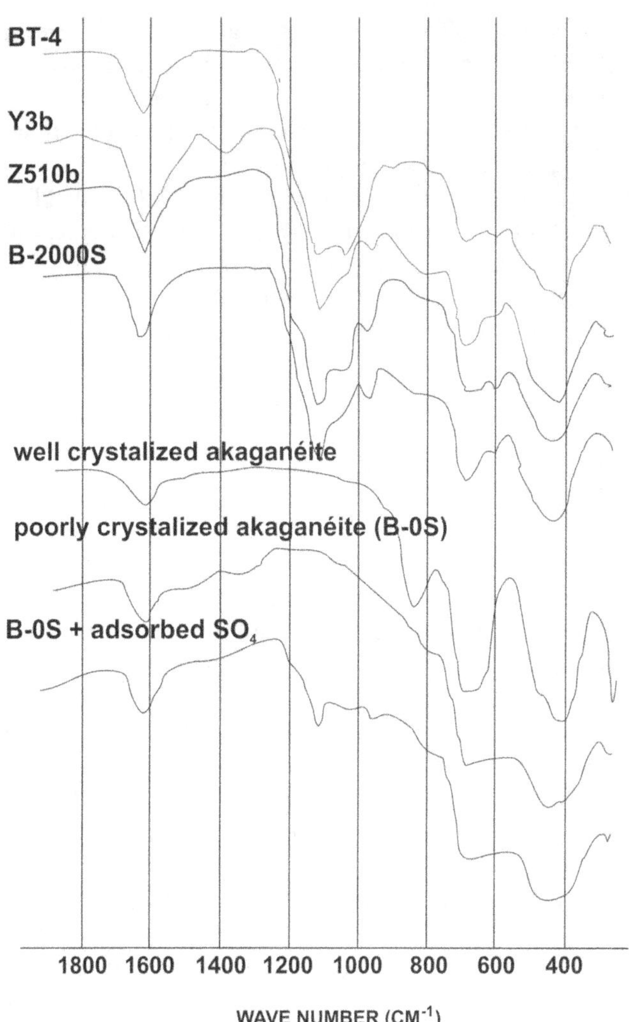

FIGURE 2.2 Infrared spectra of few natural and synthetic specimens (ref. Bigham et al., 1990). Note the similarities in peak positions between poorly crystalline akaganéite (B-0S), B-0S + adsorbed SO_4^{2-}, synthetic schwertmannite (B-2000S and Z510b), and natural schwertmannite (BT-4andY3b).Bt-4–Ohio,USA;Y3b–Finland;Z510b–biosynthesized by *Thiobacillus ferrooxidans*; and B-2000S – dialytically synthesized schwertmannite with 2000 mg L^{-1} SO_4^{2-}. 410–460/690 cm^{-1}– Fe–O stretch; 1400 cm^{-1} – COO–; 610 cm^{-1} – $\nu_4(SO_4)$; 640 cm^{-1} – OH deformation; 840 cm^{-1} – δ_{OH}; 970 cm^{-1} – $\nu_1(SO_4)$; 1110 cm^{-1} – $\nu_3(SO_4)$.

and the tunnel sulphate. The loss of these water molecules breaks some of the hydrogen bonds, which could contribute to the peak position shifts of both δ_{OH} and Fe–O stretching IR bands upon drying.

Similarity of the structure of schwertmannite to akaganéite was further confirmed using pair distribution function (PDF), XRD, and density functional theory (DFT) (Fernandez-Martinez et al., 2010). The PDFs showed that schwertmannite and akaganéite have the same structure, especially in the range of 1 to 7 Å (cf. Fig. 3 of Fernandez-Martinez et al., 2010). The first peak, at about 1.98 Å, corresponds to the Fe–O distance in Fe^{3+} octahedral coordination and is the same in both structures. The S–O bonding distance appeared at around 1.64 Å in both sulphate-doped akaganéite and schwertmannite. Peaks at approximately 3 and 3.45 Å represent Fe–Fe distances in different types of edge-sharing iron octahedra. The 3 Å peak corresponds to two Fe octahedra on the same equatorial plane, while the 3.45 Å peak corresponds to edge-sharing Fe octahedra on different equatorial planes. Peaks at 4.75, 5.45, and 6.33 Å correspond to Fe–Fe and Fe–O bonding. Although slight differences were reported, their positions and intensities were comparable. Major differences were seen between 7 and 10 Å, such as a significant decrease in the intensity of the 7.5 Å peak in the schwertmannite sample, which corresponds to correlations between iron atoms on opposite sides of the akaganéite channel. However, this was believed to be due to the presence of sulphate molecules in the structure.

PDF fitting showed that the initial akaganéite unit cell was deformed, with the β angle around 93°. Fernandez-Martinez et al. (2010) assigned a monoclinic unit cell with $a \approx 10.6$ Å, $b \approx 6.03$ Å, and $c \approx 10.6$ Å. These unit-cell parameters differ from those proposed by Bigham et al. (1994), who suggested a tetragonal unit cell with lattice parameters $a = c = 10.66$ Å and $b = 6.04$ Å. The proposed structure for the schwertmannite octahedral framework consists of a highly defective, entangled network of structural motifs. These defects may cause strains in the structure, reducing the coherent domain size to values smaller than the actual particle size. This type of strain, caused by the presence of defects or inner-sphere bidentate sulphate complexes, may be the main factor driving the deformation of the structure, along with surface energetics.

2.2.2 Alternative structural models of schwertmannite

The structure proposed at the time of its recognition, i.e., the occurrence of a tunnel structure identical to akaganéite (β-FeOOH), remains a matter of debate. Barham (1997) described schwertmannite as a class of minerals, rather than a single mineral. He suggested that it consists of Fe monomers that combine with SO_4^{2-} as a bridging ligand to form oligomers. The

SO_4^{2-} ligands can participate in ligand exchange. Barham (1997) proposed an alternative formula: $Fe_4O_4(OH)_2 \cdot AN(2/e) \cdot nH_2O$, where AN = SO_4^{2-}, CO_3^{2-}, $C_2O_4^{2-}$, NO_3^-, OH^-, ClO_4^- etc. and e is the charge of the anion. The capability to exchange ligands was demonstrated by replacing SO_4^{2-} with chromate, oxalate, and carbonate anions.

The nature of a polyphasic nano-mineral consisting of nano-crystals of hydrous ferric oxide (HFO) phases (ferrihydrite or goethite) surrounded by a sulphate-rich amorphous matrix was confirmed by HRTEM in synthetic (Hockridge et al., 2009) and natural samples (French et al., 2012), suggesting that schwertmannite should not be considered a single mineral with a repeating unit cell.

In contrast, Caraballo et al. (2013) proposed a pseudo solid-solution model for schwertmannite, rather than a defined chemical formula or true solid solutions. According to this model, two end members control the overall mineralogical behaviour, with the sulphur content playing a key role. An increase in the Fe/S molar ratio is coupled to a decrease in the sulphur content, which is linked to changes in the balance between HFO and sulphur-rich nano-domains within schwertmannite nanoparticles. Some studies, such as Hockridge et al. (2009), also suggest that schwertmannite does not have a definite crystal structure but is a combination of ordered assemblies of goethite and ferrihydrite nano-crystals.

Hockridge et al. (2009) heated 250 mL of a solution containing 400 mg of $Fe_2(SO_4)_3 \cdot xH_2O$ (0.7 g L^{-1} Fe) at 85°C for 24 hours, with regular sampling for HRTEM and XRD studies. The initial TEM images showed globular particles without surface whiskers, identified as ferrihydrite (Figure 2.3a). The lattice spacing was about 2.45 Å, close to the spacing of the (011) planes in ferrihydrite. As the reaction time increased, needle-like structures began to appear. The lattice fringes on the schwertmannite needles indicated goethite nano-crystals, with a d-spacing of ~0.4 nm, corresponding to the (101) plane of goethite. The arrangement of primary particles suggested oriented aggregates of goethite nano-crystals, twinned along the (210) plane.

After 60 minutes of reaction, the samples showed a more coherent orientation of the lattice fringes over time (Figure 2.3c). Measurements on the schwertmannite needles confirmed their structural similarity to goethite, with d-spacings of 4.25 and 2.53 Å for the (101) and (011) planes, respectively. These values closely match the d-spacings of 4.190 Å for the (101) planes and 2.522 Å for the (011) planes in goethite.

The study proposed that schwertmannite aggregates contain a central core of ferrihydrite surrounded by radiating goethite needles. It also suggested that ferrihydrite crystallites grow initially, followed by the development of goethite needles on the surface (Figure 2.3c). However, the

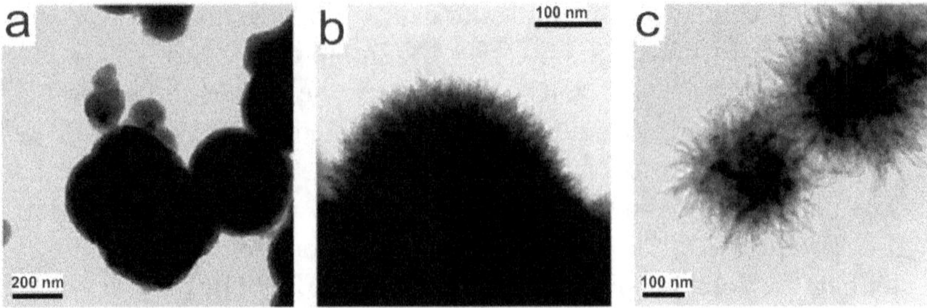

FIGURE 2.3 TEM images showing the time-resolved formation of schwertmannite aggregates from 0.7 g L^{-1} Fe^{3+} heated to 85°C after (a) 5, (b) 10, and (c) 60 minutes (ref. Hockridge et al., 2009). Note the development of surface whiskers over time.

goethite content was less than 10%, making it undetectable by XRD. Further nucleation of the needles after 60–120 minutes allowed them to grow large enough to show XRD signatures. The authors ruled out the presence of akaganéite, as it was not detected.

Schwertmannite has also been proposed to share an identical crystal structure with ferrihydrite ($5Fe_2O_3{\cdot}9H_2O$), an amorphous iron oxyhydroxide mineral phase often associated with schwertmannite. Both minerals share a common pH window of 3.0–4.5 at which they are thermodynamically stable (Loan et al., 2004). Ferrihydrite, however, has a wider stability range beyond this pH window and exhibits variable structure and crystallinity. It is a common secondary iron precipitate in acidic environments, particularly in sulphate-free or low-sulphate mine waters associated with acid mine drainage (AMD). XRD identifies two forms of ferrihydrite: 2-line and 6-line. The 2-line form is more amorphous and precipitates more quickly, resulting in a poorly ordered, poorly crystalline phase with a mixture of cubic, hexagonal, and highly disordered components. In contrast, 6-line ferrihydrite has a more hexagonal structural arrangement (Schwertmann et al., 1999; Cornell and Schwertmann, 2003).

When analysing the surface morphology of schwertmannite sampled from a mine site of the Iberian Pyrite Belt (IPB) in Spain, French et al. (2012) observed that the schwertmannite needles do not show continuous lattice fringes throughout the needles. However, in some areas, the fringes extend continuously for about 4 nm. These fringes match well with the d-spacing of the (101) plane of goethite at 0.41 nm. Other fringes recorded correspond to the intense peak of schwertmannite at 0.255 nm. Fast Fourier transformation (FFT) analysis of crystalline regions revealed that most of the d-spacing from the lattice fringes aligns with the 0.255 and 0.228 nm XRD peaks of schwertmannite and goethite. Additional lattice fringes align with the d-spacing of 0.486 nm for schwertmannite, which also matches the (200) plane of goethite. The most intense 0.255 nm peak of schwertmannite, where the majority of lattice fringes were observed, overlaps with the second (111) and third (301) most intense reflections of goethite. This overlap partially supports the argument of Hockridge et al. (2009) that goethite needles are present within schwertmannite. Similarly, Sestu et al. (2017) were able to demonstrate that traces of goethite are present in schwertmannite, but not detectable by XRD due to its negligible quantity. The wide range of variation in the solubility product K_{sp} of schwertmannite in samples studied by Caraballo et al. (2013) ($logK_{sp}$ = 5.8–39.5) is assumed to be due to the presence of goethite domains in schwertmannite, which the authors refer to as a polyphasic mineral for this reason.

It appears that schwertmannite whiskers contain a highly disordered maghemite-like structural component, which is proposed to be an intermediate phase between 2-line and 6-line ferrihydrite in terms of crystallinity and formation kinetics (Loan et al., 2004). The maghemite-like structure proposed by Loan et al. (2004) refers to research by Janney et al. (2000a, 2000b, 2001) on ferrihydrite who found that also 2-line ferrihydrite contains various structural components. Among these was a component showing similarity to the cubic unit cell of maghemite, though with slightly different atomic positions and occupancies.

In order to resolve the intermediate character of schwertmannite, Loan et al. (2005) studied the effect of the degree of supersaturation with respect to 2-line ferrihydrite as a function of pH and initial ferric iron concentration on ferrihydrite crystal growth in a $Fe_2(SO_4)_3$ solution. High supersaturation level led to the formation of 2-line ferrihydrite, whereas XRD reflections became more characteristic of schwertmannite, if supersaturation was decreased, with distinct peaks at around 41 and 74 Å 2θ, and weaker reflections between 45 and 70 Å 2θ, resembling schwertmannite. Particle size and crystallinity increased, which the authors interpreted as an intermediate phase between 2-line and 6-line ferrihydrite. TEM images of the precipitate formed at the lowest Fe(III) concentration (0.7 g L^{-1}) showed a schwertmannite-like morphology with surface whiskers (Figure 2.4). The precipitate contained approximately 25% schwertmannite. The researchers proposed that these whiskers nucleated on the surface of ferrihydrite aggregates. Indeed, mixtures of varying proportions of ferrihydrite and schwertmannite, in 3:1 ratio, produced XRD reflections identical to those of the 0.7 g L^{-1} Fe(III) solution, which contained approximately 25% schwertmannite. The difficulty to distinguish between schwertmannite and ferrihydrite is further complicated by the finding that single-area electron diffraction (SAED) of the 0.7 g L^{-1} Fe(III) only showed two diffuse rings

FIGURE 2.4 (a) TEM image of typical schwertmannite aggregates; (b) TEM image of a ferrihydrite aggregate formed from an initially high degree of Fe(III) supersaturation (c(Fe(III) = 5.2 g L^{-1} at pH 3.5, scale bar is 20 nm); and (c) TEM image of aggregates formed from a low degree of Fe(III) supersaturation (c(Fe(III) = 0.7 g L^{-1} at pH 3.5, scale bar is 0.5 mm) (ref. Loan et al., 2005). Note the schwertmannite-like morphology with surface whiskers that results from conditions of case c.

typical of 2-line ferrihydrite, while TEM and XRD analysis clearly showed surface whiskers and reflections attributable to schwertmannite.

2.3 Sulphur–Iron coordination in schwertmannite

In schwertmannite, the Fe–SO_4 double chains form 2 × 2 channels, as in akaganéite. Sulphate ions in schwertmannite are believed to form inner-sphere complexes with Fe^{3+} (Bigham et al., 1990). It appears, however, that a significant proportion of SO_4 in schwertmannite forms outer-sphere complexes as well (Majzlan and Myneni, 2005). The occurrence of outer-sphere complexes of SO_4^{2-} in schwertmannite was demonstrated by X-ray absorption near-edge structure spectroscopy (XANES), extended X-ray absorption fine structure spectroscopy (EXAFS), Fourier transform infrared spectroscopy (FTIR), and XRD studies (e.g., Boily et al., 2010; Fernandez-Martinez et al., 2010, and Wang et al., 2015). The studies conclude that two SO_4^{2-} ions are present per unit cell in the tunnel structure, with one bound as an outer-sphere complex and the other one as an inner-sphere complex presumably triggered by structural defects that allow for inner-sphere coordination. Wang et al. (2015) proposed that both of these species, i.e., inner-sphere and outer-sphere Fe–SO_4 complexes, are required to explain sulphate bonding within schwertmannite. The fraction of the inner-sphere complex decreased from 100% to ~49% as pH rose from 2 to 8, with a simultaneous increase in the fraction of outer-sphere complexes from 0% to ~50% in the case of dried schwertmannite. In wet samples, the general trend was similar, but the fraction of inner-sphere complexes decreased from ~60% to 0% and that of the outer-sphere complexes from ~40% to 100%. These findings were corroborated by sulphur K-edge XANES (Wang et al., 2015). XANES spectra always showed a pre-edge peak before the main peak. This pre-edge peak is regarded to be caused by the electronic transition from the S 1s orbital to the 3p orbital being hybridized with Fe 3d orbitals and to be indicative of SO_4 inner-sphere complexes. Compared to jarosite, in which each sulphate is coordinated to three Fe atoms, the pre-edge peaks were located at slightly lower energies. The authors concluded that this difference in energy suggests that fewer than three Fe atoms are coordinated to each sulphate in schwertmannite. Variations in the peak positions with pH and hydration degree are minimal, indicating that the type of Fe–SO_4 inner-sphere complexes is identical in schwertmannite regardless of pH.

The intensity of this pre-edge slightly decreased with increasing pH and by wetting the schwertmannite. This observation implies a decrease in the fraction of inner-sphere complexes in schwertmannite at elevated pH and the transformation of outer-sphere to inner-sphere complexes when the sample is dried (Wang et al., 2015). The shift is reversible, as comparable inner-sphere / outer-sphere distributions are achieved upon rewetting.

Overall, inner-sphere SO_4 coordination seems to be more important for dried schwertmannite than for wet schwertmannite, and vice versa, outer-sphere complexes are more important for wet samples. On the other hand, EXAFS data do not show any difference between wet and dry schwertmannites in S–O bond length, which remains constant at 1.48–1.49 Å, indicating S–O coordination is unaffected by this. Fe K-edge EXAFS demonstrated that Fe atoms are relatively more abundant in dried schwertmannites compared to wet ones, with 1.0–1.2 atoms located at a distance of 3.22–3.4 Å from the S atom in dry and 0.7–0.9 atoms at a distance of 3.23–3.26 Å from the S atom in wet schwertmannite (Wang et al, 2015). These S–Fe bond distances are close to distances determined for sulphate surface complexes on ferrihydrite (3.22 ± 0.05 Å). In addition, quantum chemical calculations predict the S–Fe distances for bidentate–binuclear and monodentate–mononuclear complexes to 3.25–3.29 Å and 3.46–3.49 Å, respectively. It is therefore assumed that the sulphate inner-sphere complexes of the dried samples are of the bidentate–binuclear type (Wang et al., 2015). The coordination numbers determined for the Fe shell (1.0–1.2) are smaller than the theoretical value of 2 expected for a bidentate–binuclear complex, which is believed to be due to the coexistence of outer-sphere complexes that cannot be detected by EXAFS. In wet schwertmannites, slightly longer Fe–S bond distances were detected that may have resulted from larger uncertainties in EXAFS fitting, because of the lower proportion of inner-sphere complexes compared to dried samples. However, consensus exists that inner-sphere complexes in both, dry and wet, schwertmannites are bidentate–binuclear.

2.4 Schwertmannite sulphate

The understanding of bonding of sulphate in schwertmannite is of paramount importance, as it influences the uptake of anionic contaminants such as AsO_3^{3-} (arsenite), SeO_4^{2-} (selenate), and PO_4^{3-} through ligand exchange with both surface-bound and structural sulphate (Paikaray et al., 2011; Antelo et al., 2013, As et al., 2024).

In schwertmannite, sulphate (SO_4^{2-}) occupies two main positions, which are tunnel sites within the mineral's structure and adsorption sites on its surface. The surface-bound sulphate is more easily desorbed as the pH increases (Paikaray and Peiffer, 2010), which in turn raises the proportion of sulphate in tunnel sites over time as schwertmannite is exposed to neutral-to-alkaline aqueous media.

Evidence for SO_4 coordination changes upon increase in pH is recorded in X-ray diffractograms. Well distinct, yet broad, eight diffractogram peaks are obtained from freshly synthesized schwertmannite at pH ~ 3.2–3.5

(Figure 2.5). However, these peak positions shift as pH increases towards alkaline domain (pH ~ 8.0), e.g., peak positions at (301), (225), and (040) shift towards lower d-spacing and the (200) peak to a higher d-spacing together with lowering of the intensities of the (200) and the (301) peaks (Wang et al., 2015). Traces of goethite also form at such alkaline conditions of pH 8 with a low intensity of the (110) XRD reflection. Such peak shifting relative to the original schwertmannite is attributed to structural modifications within schwertmannite before its transformation to goethite coupled with SO_4^{2-} desorption from both surface sites and structural bonding sites of schwertmannite. The shifting of the diffractogram peaks upon increase of pH is accompanied by substantial SO_4^{2-} release and the lowering of IR band intensities (Wang et al., 2015).

While there is consensus on the important role of sulphate in schwertmannite, there is some debate regarding its exact positioning within the structure. Waychunas et al. (2001) suggested that sulphate ions might occupy both inner-sphere and outer-sphere positions on the surface of schwertmannite crystallites, as well as within defective regions of the structure. This variation in sulphate positioning may further influence schwertmannite's structural integrity and its ability to interact with environmental contaminants.

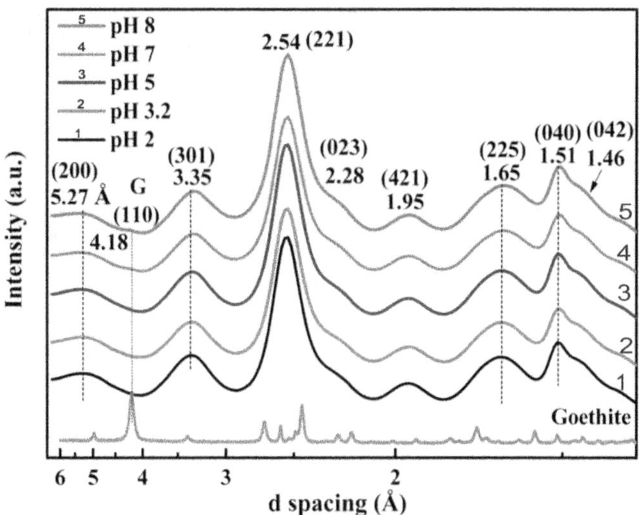

FIGURE 2.5 X-ray diffraction patterns for the air-dried schwertmannite samples after incubation for 24 hours at various pH values in 0.05 M $NaNO_3$ solution (G = goethite, included as a reference) (ref. Wang et al., 2015). Note the peak shift upon increase in pH and appearance of low-intensity goethite peaks at pH 8.0.

The sulphate (SO_4^{2-}) vibrational modes in schwertmannite exhibit notable changes in response to variations in pH and temperature. In freshly synthesized schwertmannite, the $v_3(SO_4)$ vibration (asymmetric stretching) is located at a wavenumber of ~1120 cm^{-1} with shoulders around 1030 and 1190 cm^{-1}, while the $v_1(SO_4)$ vibration (symmetric stretching) appears near ~985 cm^{-1} and the $v_4(SO_4)$ vibration (bending mode) is found around 610 cm^{-1} (Bigham et al., 1990). The positions of these vibrations may slightly vary depending on the specific sample. Upon changes in pH, Wang et al. (2015) observed shifts in band positions at 1104 cm^{-1} (with shoulders at 1040 and 1170 cm^{-1} for $v_3(SO_4)$), 980 cm^{-1} for $v_1(SO_4)$, and 608 cm^{-1} for $v_4(SO_4)$. These shifts were attributed to inner-sphere SO_4^{2-} complexes, which suggest that the sulphate ions are directly coordinated to the Fe atoms within schwertmannite's structure.

As the pH increases from acidic (~3.1) to alkaline conditions (~9.0), the intensity of the $v_3(SO_4)$, $v_1(SO_4)$, and $v_4(SO_4)$ vibrations decreases. These observations were linked to changes in the speciation of SO_4^{2-} in schwertmannite under different pH conditions. As pH becomes more alkaline, schwertmannite gradually transforms into a mixture of goethite and schwertmannite, leading to a desorptive loss of structural sulphate. A comparison between dry and wet schwertmannite samples revealed that the $v_3(SO_4)$ vibration band exhibits greater splitting in dry schwertmannite (Wang et al., 2015) confirming that dry samples contain relatively more inner-sphere sulphate complexes than wet samples, where the presence of water may alter the bonding environment of sulphate within the structure. Regardless of the proportion of the surface sulphate, the same S–Fe interatomic distance is observed, indicating that surface sulphate has similar local atomic environment as the tunnel sulphate.

In an attempt to identify sulphate components in schwertmannite and their pH dependent behaviour, Boily et al. (2010) classified SO_4^{2-} vibrations based on patterns observed in sorption studies. Bands at 1120, 1070, 1033, and 990 cm^{-1} were grouped as component I and are similar to SO_4 adsorbed at hematite and goethite and schwertmannite surfaces in the presence of H_2O. These bands contribute to ~50% of the spectra at low pH conditions making it the predominant spectral component of schwertmannite (Boily et al., 2010). Component II comprises bands at 1210, 1180, 1130, 1033, 972, and 966 cm^{-1}, mostly v_1 splitting, which are comparable to spectra of SO_4 adsorbed to the hematite surface at circum-neutral pH, at which conditions they make up ~75% of spectral intensities of a schwertmannite sample. These spectra have distinctly lower intensities at acidic or alkaline conditions. Component III at 1108 and 972 cm^{-1} is comparable to SO_4 adsorbed onto goethite and constitutes ~75% of spectral intensities at alkaline conditions, with lower intensities under circum-neutral conditions. Component III bands are resembling the spectra of the tetrahedral aqueous SO_4^{2-} ion that corresponds to relatively undistorted hydrogen-bonded complexes and presumably can be attributed to outer-sphere complexes. Component I is regarded to be a mixture of SO_4^{2-} complexes with Fe^{3+} and hydrogen, both inner and outer spheres, and matches monodentate (1137–1143 cm^{-1}), bidentate mononuclear (1043–1070 cm^{-1}), and bidentate binuclear (1006–1032 cm^{-1}) SO_4 complexes in Fe^{3+}–SO_4 bonding (Paul et al., 2005). Component II was assigned to the protonation of SO_4^{2-} on mineral surfaces.

2.5 Hydroxyl groups in schwertmannite

As discussed in the previous section, OH groups are structurally associated with Fe atoms in normal octahedral coordination with O and OH as $FeO_3(OH)_3$, and with partial replacement by SO_4 to yield $FeO_3(OH)_2O$–SO_3.

Boily et al. (2010) reported the presence of an IR band corresponding to OH deformation vibrations at 860 cm^{-1} (δ_{OH}) and stretch vibrations (ν_{OH}) at 3300 cm^{-1} for schwertmannite under acidic pH conditions (pH < 4). Once exposed to circum-neutral or alkaline pH conditions, schwertmannite was partly transformed to goethite and exhibited slight shifts in δ_{OH} (~790 and ~895 cm^{-1}) and ν_{OH} (~3150 cm^{-1}). The water content becomes marginally lower due to goethite formation. Wang et al. (2015) also assigned ~840 cm^{-1} to δ_{OH}-vibrations and found similar reductions in its intensity with increasing pH, along with splitting into ~892 cm^{-1} (δ_{OH}) and ~795 cm^{-1} (γ_{OH}) goethite bands. The δ_{OH} (~840 cm^{-1}) and Fe–O (~685 cm^{-1}) vibration bands shift to lower wavenumbers (~786 and ~665 cm^{-1}) for wet schwertmannites compared to their dry counterparts, indicating stronger hydrogen bonding in the structure under wet conditions than under dry conditions.

2.6 Contaminants' effect on schwertmannite structure

Schwertmannite precipitating environments such as AMD, mine pit lakes, or acid sulphate soils (ASS) are often enriched in contaminants and may be involved in the precipitation of schwertmannite form Fe^{3+}- and SO_4^{2-}-rich aqueous solutions thereby affecting its structure significantly.

Waychunas et al. (1995) studied the bonding behaviour of selenate (Se(VI)) and arsenate (As(V)) in schwertmannite synthesized in the presence of these oxoanions using X-ray absorption spectroscopy. In selenate-substituted schwertmannite, no changes were observed in the Se or Fe near-edge EXAFS or in the structural function when the Se concentration changed. The Se–O bond distance was measured at 1.64 Å, with a coordination number of 4.3, consistent with undistorted selenite tetrahedra. Two Se–Fe distances were reported: 1 Fe neighbour at 3.23 Å and 5 Fe neighbours at 3.40 Å. The average Se–Fe distance was slightly longer than expected for a bidentate sorption complex (3.37 Å versus ~3.23 Å), but shorter than what would be expected for tunnel occupation. The authors concluded that selenate replaces sulphate in schwertmannite, likely occupying tunnel sites and also adsorbing onto the crystallite surfaces. Hence, selenate substituted directly for sulphate within the tunnels of the schwertmannite structure and also adsorbed onto the surface without disrupting the structure.

In contrast, arsenate primarily adsorbed onto the surface of schwertmannite. Moreover, the structure was destabilized and its growth was inhibited. For arsenate, the As–O shell had 4.5 oxygen neighbours at 1.678 Å, and the As–Fe shell had 2.5 iron neighbours at 3.28 Å which is indicative for the formation of a bidentate sorption complex on schwertmannite, similar to surface complexes of arsenate formed on the surface of akaganéite and goethite (Waychunas et al. (1995). Unfortunately, no concentration was given for the As concentrations at which the synthesis was made in the study by Waychunas et al. (1995) and the effects of arsenate seem to be concentration dependent. Schwertmannite, synthesized in solutions containing arsenate in addition to sulphate, was enriched by up to 10.3 wt% arsenate without structural changes detectable by powder XRD (Regenspurg and Peiffer, 2005). In contrast to arsenate and similar to selenate, a total substitution of sulphate by chromate was possible in sulphate-free solutions. Thereby, the chromate content in schwertmannite could reach 15.3 wt%.

Similarly, varying As(III)/Fe(III) molar ratios (0.0–0.8) during schwertmannite precipitation did not significantly alter its local structure as demonstrated by Fe K-edge EXAFS spectroscopy. However, there was a marked decrease in the intensity of the second-neighbour peak in the Fourier transform of the Fe K-edge XANES spectra, which was attributed to particle size variation (Paikaray et al., 2011). Moreover, XRD peaks broadened with increased As(III) loading, which is indicative of a reduction of particle size. Other than As(V), the presence of As(III) may increase structural disorder in schwertmannite rather than reducing crystallite size.

2.7 Iron and sulphur variations in schwertmannite

Environments in which schwertmannite occurs are shaped by a wide range of pH and pE (a measure of redox potential) values. A comprehensive study by Caraballo et al. (2013), which compiled data from

30 schwertmannite samples from different countries, found that the mineral can precipitate within a pH range of 1.93 to 4.71 and a pE range of 8.50 to 13.70. Moreover, schwertmannite formation is also influenced by the concentrations of Fe^{3+} and SO_4^{2-} ions in the environment, with studies showing a wide range of favourable conditions for its precipitation. Fe^{3+} concentrations in environments where schwertmannite occurs can vary from 30 μM to 19 mM, while SO_4^{2-} concentrations range from 7 to 97 mM (Caraballo et al., 2013). Yet, the molar ratio of Fe/S in schwertmannite is considered to lie within a relatively narrow range of 4.0 to 8.0 (Bigham and Nordstrom, 2000), although it may vary between 3.8 and 15.5, with the most common molar ratio between 4 and 6 (Caraballo et al., 2013). A linear regression was established for the Fe and S content in schwertmannite ($R^2 = 0.69$) based on 46 natural schwertmannite samples (Caraballo et al., 2013) (Eq. 2.1).

$$c(\text{Fe}) = 4.41 + 1.94c(\text{S})\left[\text{mmol g}^{-1}\right] \tag{Eq. 2.1}$$

The stoichiometry of schwertmannite dissolution (Eq. 2.2, Bigham et al., 1996) predicts that changes in the sulphate content (y) directly affect the hydroxide content (OH^-), i.e., with any SO_4^{2-} ion taken up the number of OH groups is reduced by a factor of 2. Any alteration in the sulphate content impacts the overall molar mass of schwertmannite, but for typical values of y (ranging from 0.5 to 2.0), these mass changes are generally negligible, usually less than 1%.

$$\begin{aligned} Fe_8O_8(OH)_{8-2y}(SO_4)_y + (24-2y)H^+ &= \\ 8Fe^{3+} + ySO_4^{2-} + (16-2y)H_2O \end{aligned} \tag{Eq. 2.2}$$

2.8 Solubility of schwertmannite

Based on the stoichiometry in Eq. 2.1, Bigham et al. (1996) proposed an ion activity product (IAP) for the dissolution of schwertmannite (Eq. 2.3) as:

$$\log \text{IAP} = 8\log a_{Fe^{3+}} + y\log a_{SO_4^{2-}} + (24-2y)\text{pH} \tag{Eq. 2.3}$$

with a_x being the activity of the species x based on measurements in both synthetic samples and in aqueous systems containing schwertmannite. By means of chemical modelling of measured data, they proposed a solubility window of schwertmannite (SHM) with

$$\log \text{IAP}_{SHM} = 18.0 \pm 2.5 \tag{Eq. 2.4}$$

that allowed to construct the stability field of schwertmannite and associated secondary iron precipitates, i.e., ferrihydrite, jarosite, and goethite by means of the computer code MINTEQ A (Figure 2.6). In the simulations, the ferrihydrite–schwertmannite boundary could be reproduced. However, the simulation is quite sensitive for the simulation of jarosite. Depending on the logK_{sp} value for jarosite as discussed in Bigham et al. (1996), i.e., reported values between −12.51 and −9.21 for the following dissolution reaction (Eq. 2.5).

$$KF_3(SO_4)_2(OH)_6 + 6H^+ = K^+ + 3Fe^{3+} + 2SO_4{}^{2-} \cdot 6H_2O \qquad \text{(Eq. 2.5)}$$

The jarosite stability field may obscure the schwertmannite stability field. Occurrence of a jarosite higher solubility would result in a narrower schwertmannite stability field. Overall, these simulations suggest a variable stability field of schwertmannite depending on the crystallinity, particle size, and differences in free energy of the various Fe(III)-phases involved (Figure 2.6).

The value for the IAP was widely used for thermodynamic calculations related to the occurrence of schwertmannite. As an example, Regenspurg et al. (2004) demonstrated that acidic mining lakes were at redox equilibrium with Fe^{2+}- and schwertmannite-containing lake sediments. Little deviation from this value was observed by Burton et al. (2008) who calculated an IAP_{SHM} value of $10^{19\pm2}$ for pore waters of coastal acidic soils of eastern Australia containing schwertmannite.

In contrast, Kawano and Tomita (2001) derived an IAP for schwertmannite from the composition of a spring site draining pyritic host rock to be 7.06 ± 0.09 at 30°C based on the following stoichiometry (Eqs. 2.6 and 2.7). This further suggests that nature of sample determines the solubility product of schwertmannite, especially the $SO_4{}^{2-}$ content

$$Fe_8O_8(OH)_{5.90}(SO_4)_{1.05} + 21.9H^+ \rightarrow 8Fe^{3+} + 1.05SO_4{}^{2-} + 13.9H_2O \qquad \text{(Eq. 2.6)}$$

$$\log IAP_{SHM} = \log \frac{\left[Fe^{3+}\right]^8 \left[SO_4{}^{2-}\right]^{1.05}}{\left[H^+\right]^{21.9}} \qquad \text{(Eq. 2.7)}$$

Yu et al. (2002) demonstrated very little or no structural OH to occur in the schwertmannite samples they were synthesizing at pH ~ 2 and proposed an alternative stoichiometry as $Fe_2O_{3-x}(SO_4)_x \cdot nH_2O$; x = 0.41–0.49 and n = 1.51–2.81. The logIAP values estimated from the activities of the corresponding species in their synthesis solution by means of the computer

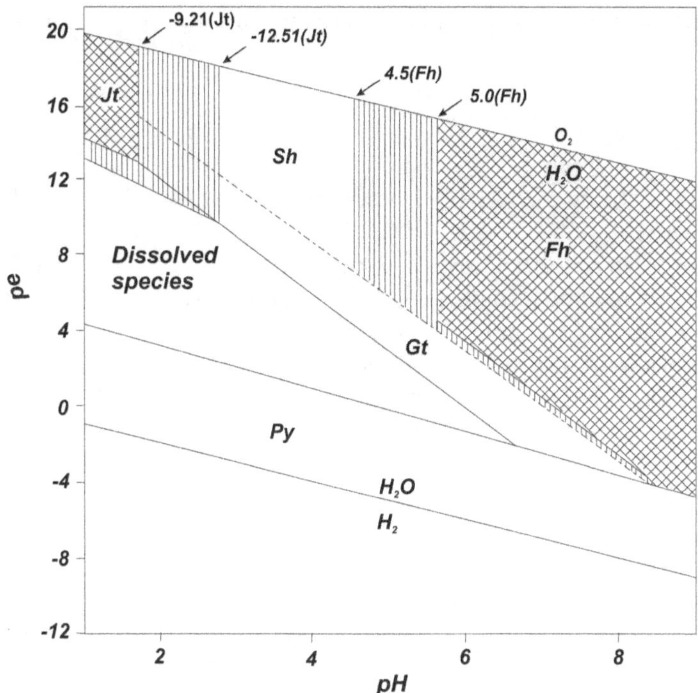

FIGURE 2.6 pE–pH diagram for Fe–S–O–K–H system at 25°C. pE = Eh/59.2; total log activities of Fe^{2+} = −3.47; Fe^{3+} = −3.36 or −2.27; SO_4^{2-} = -2.32; K^+ = −3.78; $\log a_{Fe^{3+}} = 1.40 - 3pH$ (goethite); $\log a_{Fe^{3+}} = -0.30 - 2pH$ (K-jarosite); $\log a_{Fe^{3+}} = 5.0 - 3pH$ (ferrihydrite); $\log a_{Fe^{3+}} = 2.67 - 2.6pH$ (schwertmannite). Line equations are: Gt (pE = 17.9–3 pH); Jt (pE = 16.21–2 pH); Fh (pE = 21.50–3 pH); Sh (pE = 19.22–2.6 pH); and Py (pE = 5.39–1.14 pH). Fields of metastability are shown by dashed lines. Single-hatched areas demonstrate expansion of K-jarosite and ferrihydrite fields if lower $\log K_{sp}$s are selected (ref. Bigham et al., 1996).

program PHREEQC were 2.01 ± 0.30. This value was comparable to that proposed by Yu et al. (1999) of $\log IAP_{SHM}$ = 2.68 between 10°C and 15°C for a similar stoichiometry of stream waters affected by AMD (Korea), which the authors explained by a lower solubility of schwertmannite than the value suggested by Bigham et al. (1996).

The authors proposed a pH-dependent ion exchange equilibrium between schwertmannite and ferrihydrite between SO_4^{2-} and OH^- (Eq. 2.8).

$$Fe_2O_{3-x}(SO_4)_x + (x + 0.5y)H_2O = Fe_2O_{3-0.5y}(OH)_y + xSO_4^{2-} + 2xH^+ \qquad \text{(Eq. 2.8)}$$

that depends on both, pH and the activity of SO_4^{2-}, expressed as pSO_4 (Eq. 2.9)

$$pSO_4 = -2pH + \frac{(pIAP_{SHM} - pK_{sp,Fh})}{x} \quad \text{(Eq. 2.9)}$$

with $pK_{sp,Fh}$ being the negative decadic logarithm of solubility product for ferrihydrite.

Hence, determining a precise dissolution reaction and solubility product for schwertmannite is challenging due to its compositional variability, temporal instability, and ability to adsorb species onto its surface. The application of established logIAP values from Bigham et al. (1996), Yu et al. (1999), and Kawano and Tomita (2001) (18 ± 2.5, 10.5 ± 2.5, and 7.06 ± 0.09, respectively) often results in contrasting saturation behaviours. For example, Santofimia et al. (2015) observed this inconsistency in the Nuestra Señora del Carmen (NSC) mine pit lake, IPB, Spain, where chemical data and saturation indices (SI) suggested supersaturation for schwertmannite, even though it was not observed in the lake, and vice versa under opposing conditions. The authors recalculated the logIAP values using the geochemistry of the pit lake and proposed a value of 10.5–11.5, which aligns closely with the value suggested by Yu et al. (1999).

In an attempt to generalize the various observations, Caraballo et al. (2013) established a predominance diagram for 30 schwertmannite samples collected from various parts of the world, including the USA, Germany, Canada, and Japan, by plotting measured pE and pH values into the pE–pH diagram. The calculated IAP values ranged from 5.8 to 39.5 (Caraballo et al., 2013). The researchers then combined this predominance diagram with a pE–pH diagram for stable iron minerals like goethite, hematite, and pyrite to define the boundary of the schwertmannite predominance range (Figure 2.7).

In the generated pE–pH diagram, the stability field for goethite is quite small, partially covering the stability fields of ferrihydrite and schwertmannite. This limited stability field for goethite is thought to be due to the coexistence of both ferrihydrite and schwertmannite, which act as precursors to goethite formation. Direct precipitation of goethite in this pH range (pH ~ 2.0–5.0) is considered unlikely. Schwertmannite exhibits a very wide stability field compared to that proposed by Bigham et al. (1996) (cf. Figure 2.6) because of the large number of environments from which schwertmannite samples were collected. The authors proposed that the logIAP value for schwertmannite should not be treated as a single, fixed value. Rather it varies with the sulphur content of schwertmannite as predicted by the statistical relationship in Eq. 2.10 ($R^2 = 0.68$).

$$\log IAP = 40.40 - 15.10x(SO)_4 \quad \text{(Eq. 2.10)}$$

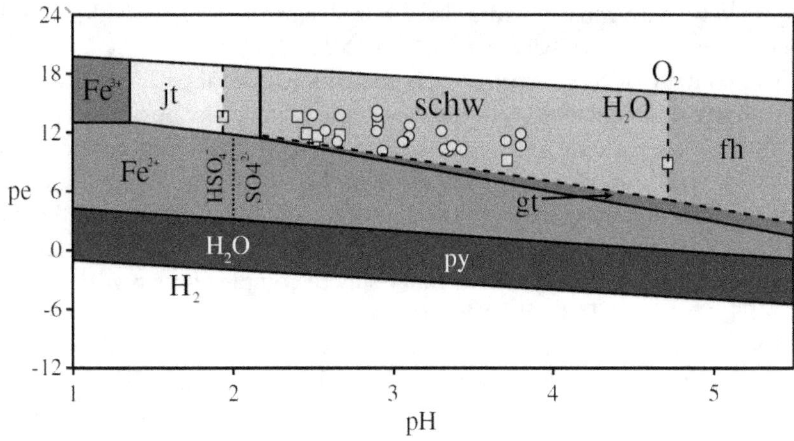

FIGURE 2.7 pE–pH diagram for stable and metastable iron mineral phases (ref. Caraballo et al., 2013). Abbreviations: Fe^{3+}, Fe^{2+}, SO_4^{2-}, HSO_4^-, and H_2O = dissolved species; H_2 and O_2 = gas species; jt = jarosite; schw = schwertmannite; gt = goethite; fh = ferrihydrite; and py = pyrite. Dashed lines are employed to delimit the predominance fields of the metastable phases (schw and fh). Square symbols correspond to field samples from the Iberian Pyrite Belt, and circle symbols stand for natural samples that were compiled from literature data by the authors.

with $x(SO_4)$ being the stoichiometric sulphate content in schwertmannite per 8 mol Fe.

Overall, the thorough analysis by Caraballo et al. (2013) clearly covers the broad range of $\log IAP_{SHM}$ values reported in earlier studies, once the composition of the schwertmannite sample is considered. The authors are highlighting the model's utility for predicting behaviour in various earth and planetary environments, such as Martian surfaces, where the necessary information to calculate logIAP may be lacking.

2.9 Summary

Because of its amorphous nature, fine size, and poor crystallinity, determination of the structure of schwertmannite remains a challenge until today. Meta-analysis of 30 schwertmannite samples from different mine sites across many countries suggests that schwertmannite can precipitate across a pH range from 1.93 to 4.71 and a pE ranging from 8.50 to 13.70, with an average pH of 3.02 ± 0.56 and a mean pE of 11.37 ± 1.37. Dissolved Fe^{3+} and SO_4^{2-} are two important constituents for schwertmannite formation ranging in concentrations from 30 μM to 19 mM and 7 to 97 mM, respectively, with most favourable Fe/S molar ratios between 4.0 and 8.0, which could even vary between 3.8 and 15.5.

The content of SO_4^{2-} within schwertmannite is crucial for controlling its stability and reactivity. Sulphate occupies two main positions: tunnel sites within the mineral's structure and adsorption sites on its surface. The surface-bound sulphate is more easily desorbed as the pH increases, while tunnel SO_4^{2-} influences the uptake of anionic contaminants through ligand exchange process. Sulphate ions in schwertmannite form inner-sphere complexes with Fe(III) as well as outer-sphere complexes as revealed from XANES, EXAFS, FTIR, and XRD studies.

The schwertmannite structure was first proposed to be similar to that of akaganéite, but the characteristic (110) XRD reflection of akaganéite at 0.747 nm is missing in schwertmannite. This suggests a structural difference between the two minerals. The role of SO_4^{2-} in schwertmannite's structure is thought to be analogous to that of Cl^- in akaganéite. In schwertmannite, SO_4^{2-} shares two oxygen atoms (O^{2-}, OH^-) along the walls of narrow tunnels, while the other two oxygen atoms are situated along the centre of the tunnels. The tunnel in akaganéite has a cross-section of approximately 4 × 4 Å, which is large enough to accommodate Cl^-, OH^-, or water, but too small for SO_4^{2-}. To fit SO_4^{2-} into these structural cavities, the sulphate ions share oxygen atoms with surrounding iron atoms, causing significant distortion in the structure. This distortion likely accounts for the absence of the akaganéite (110) reflection in schwertmannite. The structural similarity between schwertmannite and akaganéite was confirmed by many research studies using advanced analytical techniques such as XANES, EXAFS, WAXS, Mössbauer, and statistical interpolations like PDF and DFT.

Many studies have disagreed with the akaganéite structural resemblance of schwertmannite and proposed multiple views. Some studies proposed schwertmannite as a class of minerals, rather than a single mineral, and proposed the alternative formula $Fe_4O_4(OH)_2 \cdot AN(2/e) \cdot nH_2O$, where AN = SO_4^{2-}, CO_3^{2-}, $C_2O_4^{2-}$, NO_3^-, OH^-, ClO_4^-, etc., and e is the charge on the anion. Schwertmannite should not be considered a single mineral with a repeating unit cell, rather it should be viewed as a polyphasic nano-mineral consisting of nano-crystals of HFO phases (ferrihydrite or goethite) surrounded by a sulphate-rich amorphous matrix. Few studies proposed that schwertmannite aggregates contain a central core of ferrihydrite surrounded by radiating goethite needles where ferrihydrite crystallites grow initially followed by the development of goethite needles on the surface. Other studies proposed that a fraction of goethite is present in schwertmannite, but not detected by XRD due to its negligible quantity. Highly disordered maghemite-like structural component in schwertmannite whiskers is detected in few studies. The presence of contaminants, particularly those that may occupy the structure, e.g., As(III)/As(V), may destabilize the structure and inhibit its growth.

The solubility of schwertmannite is found to be variable depending on its nature, most importantly on the SO_4^{2-} content. The IAPs determined so far vary significantly based on geochemical variation in schwertmannite precipitating environments and the chemical composition. Compilation of 30 schwertmannites from different countries and different environmental settings resulted in a logIAP variation between 5.8 and 39.5. Hence, determining a precise dissolution reaction and solubility product for schwertmannite is challenging due to its compositional variability, temporal instability, and ability to adsorb species onto its surface.

References

Antelo J, Fiol S, Gondar D, Perez C, Lopez R and Arce F (2013). Cu(II) incorporation to schwertmannite: Effect on stability and reactivity under AMD conditions. *Geochimica et Cosmochimica Acta* 119: 149–163.

As K, Peiffer S, Uhuegbue PO, Joshi P, Kappler A, Marouane B and Hockmann K (2024). Sulfate affinity controls phosphate sorption and the proto-transformation of schwertmannite. *Chemical Geology* 653: 122043.

Barham RJ (1997). Schwertmannite: A unique mineral, contains a replaceable ligand, transforms to jarosites, hematites, and/or basic iron sulphate. *Journal of Materials Research* 12: 2751–2758.

Bigham JM, Carlson L and Murad E (1994). Schwertmannite, a new iron oxyhydroxysulfate from Pyhasalmi, Finland, and other localities. *Mineralogical Magazine* 393: 641–648.

Bigham JM and Nordstrom DK (2000). Iron and aluminum hydroxysulfates from acid sulphate waters. In: Alpers CN, Jambor JL, Nordstrom DK (Eds), *Reviews in mineralogy and geochemistry, sulphate minerals: Crystallography, geochemistry, and environmental significance*, vol. 40. The Mineralogical Society of America, pp. 351–403.

Bigham JM, Schwertmann U, Carlson L and Murad E (1990). A poorly crystallized oxyhydroxysulfate of iron formed by bacterial oxidation of Fe(II) in acid mine waters. *Geochimica et Cosmochimica Acta* 54: 2743–2758.

Bigham JM, Schwertmann U, Tranina SJ, Winland RL and Wolf M (1996). Schwertmannite and the chemical modeling of iron in acid sulphate waters. *GeochimicaetCosmochimicaActa* 60: 2111–2121.

Boily JF, Gassman PL, Peretyazhko T, Szanyi J and Zachara JM (2010). FTIR spectral components of schwertmannite. *Environmental Science and Technology* 44: 1185–1190.

Brady KS, Bigham JM, Jaynes WF and Logan TJ (1986). Influence of sulphate on Fe-oxide formation: Comparisons with a stream receiving acid mine drainage. *Clays and Clay Minerals* 34: 266–274.

Burton ED, Bush RT, Sullivan LA and Mitchell DRG (2008). Schwertmannite transformation to goethite via the Fe(II) pathway: reaction rates and implications for iron-sulfide formation. *Geochimica et Cosmochimica Acta* 72: 4551–4564.

Caraballo MA, Rimstidt JD, Macías F, Nieto JM and Hochella Jr. MF (2013). Metastability, nanocrystallinity and pseudo-solid solution effects on the understanding of schwertmannite solubility. *Chemical Geology* 360–361: 22–31.

Cornell RM and Schwertmann U (2003). *The iron oxides: Structure, properties, reactions, occurences and uses.* 2nd Ed. Wiley-VCH Verlag GmbH & Co. KGaA.

Fernandez-Martinez A, Timon V, Roman-Ross G, Cuello GJ, Daniels JE and Ayora C (2010). The structure of schwertmannite, a nanocrystalline iron oxyhydroxysulfate. *American Mineralogists* 95: 1312–1322.

French RA, Caraballo MA, Kim B, Rimstidt JD, Murayama M and Hochella Jr. MF (2012). The enigmatic iron oxyhydroxysulfate nanomineral schwertmannite: Morphology, structure, and composition. *American Mineralogist* 97: 1469–1482.

Hockridge JG, Jones F, Loan M and Richmond WR (2009). An electron microscopy study of the crystal growth of schwertmannite needles through oriented aggregation of goethite nanocrystals. *Journal of Crystal Growth* 311: 3876–3882.

Janney DE, Cowley JM and Buseck PR (2000a). Structure of synthetic 2-line ferrihydrite by electron nanodiffraction. *American Mineralogist* 85: 1180–1187.

Janney DE, Cowley JM and Buseck PR (2000b). Transmission electron microscopy of synthetic 2- and 6-line ferrihydrite.*Clays and Clay Minerals*, 48: 111–119.

Janney DE, Cowley JM and Buseck PR (2001). Structure of synthetic 6-line ferrihydrite by electron nanodiffraction. *American Mineralogist* 86: 327–335.

Kawano M and Tomita K (2001). Geochemical modeling of bacterially induced mineralization of schwertmannite and jarosite in sulfuric acid spring water. *American Mineralogist* 86: 1156–1165.

Loan M, Hart RD, Cowley JM and Parkinson GM (2004). Evidence on the structure of synthetic schwertmannite. *American Mineralogist* 89: 1735–1742.

Loan M, Richmond WR and Parkinson GM (2005). On the crystal growth of nanoscale schwertmannite. *Journal of Crystal Growth* 275: e1875–e1881.

Majzlan J and Myneni SCB (2005). Speciation of iron and sulphate in acid waters: Aqueous clusters to mineral precipitates. *Environmental Science and Technology* 39: 188–194.

Mazzetti L and Thistlethwaite PJ (2002). Raman spectra and thermal transformations of ferrihydrite and schwertmannite. *Journal of Raman Spectroscopy* 33: 104–111.

Paikaray S, Göttlicher J and Peiffer S (2011). Removal of As(III) from acidic waters using schwertmannite: Surface speciation and effect of synthesis pathway. *Chemical Geology* 283: 134–142.

Paikaray S and Peiffer S (2010). Dissolution kinetics of sulphate from schwertmannite under variable pH conditions. *Mine Water and the Environment* 29: 263–269.

Paul KW, Borda MJ, Kubicki JD and Sparks DL (2005). Effect of dehydration on sulphate coordination and speciation at the Fe-(hydr)oxide-water interface: A molecular orbital/density functional theory and Fourier transform infrared spectroscopic investigation. *Langmuir* 21: 11071–11078.

Regenspurg S, Brand A and Peiffer S (2004). Formation and stability of schwertmannite in acidic mining lakes. *Geochimica et Cosmochimica Acta* 68: 1185–1197.

Regenspurg S and Peiffer S (2005). Arsenate and chromate incorporation in schwertmannite. *Applied Geochemistry* 20: 1226–1239.

Santofimia E, López-Pamo E and Montero E (2015). Selective precipitation of schwertmannite in a stratified acidic pit lake of Iberian Pyrite Belt. *Mineralogical Magazine* 79: 497–513.

Schwertmann U, Friedll J and Stanjek H (1999). From Fe(III) ions to ferrihydrite and then to hematite. *Journal of Colloid and Interface Science* 209: 215–223.

Sestu M, Navarra G, Carrero S, Valvidares SM, Aquilanti G, Pérez-Lopez R and Fernandez-Martinez A (2017). Whole-nanoparticle atomistic modeling of the schwertmannite structure from total scattering data. *Journal of Applied Crystallography* 50: 1617–1626.

Wang X, Gu C, Feng X and Zhu M (2015). Sulphate local coordination environment in schwertmannite. *Environmental Science and Technology* 49: 10440–10448.

Waychunas GA, Myneni SCB, Traina SJ, Bigham JM, Fuller CC and Davis JA (2001). Reanalysis of the schwertmannite structure and the incorporation of SO_4^{2-} groups: An IR, XAS, WAXS and simulation study. In: Bodnar RJ (Ed), *Abstracts of the 11th Goldschmidt conference.* Varginia, p. 3849.

Waychunas GA, Xu N, Fuller CC, Davis JA and Bigham JM (1995). XAS study of AsO_4^{3-} and SeO_4^{2-} substituted schwertmannites. *Physica B* 208 & 209: 481–483.

Yu J, Heo B, Choi I, Cho J and Chang H (1999). Apparent solubilities of schwertmannite and ferrihydrite in natural stream waters polluted by mine drainage. *Geochimica et Cosmochimica Acta* 63: 3407–3416.

Yu J, Park M and Kim J (2002). Solubilities of synthetic schwertmannite and ferrihydrite. *Geochemical Journal* 36: 119–132.

3

SYNTHESIS OF SCHWERTMANNITE

3.1 Introduction

Schwertmannite can be synthesized in the laboratory by controlling iron and sulphate concentrations under various conditions. Laboratory synthesis is often necessary for research to better understand schwertmannite's structure, ion exchange properties, contaminant uptake, and physico-chemical characteristics. Since pure forms of naturally occurring schwertmannite are rarely available, researchers rely on laboratory methods for its preparation. Different laboratories have produced schwertmannite samples with diverse chemical compositions and incorporated trace elements.

Schwertmannite can be synthesized using both abiotic and biotic methods. Abiotic synthesis involves the rapid hydrolysis of ferric nitrate or chloride solutions containing sulphate (Brady et al., 1986; Bigham et al., 1990). Biotic synthesis, on the other hand, involves the bacterial oxidation of acidic ferrous sulphate solutions using strains of *Thiobacillus ferrooxidans* (Lazaroff et al., 1982, 1985; Bigham et al., 1990) or *Acidithiobacillus ferrooxidans* (Wang et al., 2006; Qiao et al., 2017), typically in batch or dynamic fermentation systems. In these biotic methods, the oxidation of Fe^{2+} is mediated by bacteria, but mineralization occurs outside the cells, meaning that mineral formation is controlled by pH and concentrations of SO_4^{2-} and Fe^{3+}. The time required for synthesis varies widely, from a few minutes to more than a month.

The synthesis of pure schwertmannite requires iron and sulphur concentrations in appropriate proportions to maintain its stoichiometry. The iron content typically ranges between 7.7 and 10.2 mmol g^{-1} (Bigham et al.,

DOI: 10.1201/9781003583875-3

1990; Regenspurg et al., 2004), while the sulphur content ranges from 1.4 to 2.1 mmol g^{-1}, with an Fe/S molar ratio between 4.6 and 8 in schwertmannite (Bigham et al., 1990, 1994, 1996). As a result, methods used for schwertmannite synthesis closely adhere to this Fe/S molar ratio by adjusting the concentrations of iron and sulphur in the solution accordingly.

Essentially, three methods are proposed for the synthesis of schwertmannite: (1) rapid oxidation of Fe^{2+} in the presence of SO_4^{2-} using hydrogen peroxide (H_2O_2); (2) hydrolysis of Fe^{3+} in the presence of SO_4^{2-}, followed by dialysis of the resulting precipitate; and (3) bacterial oxidation of Fe^{2+}. The first two methods are abiotic, while the third one involves Ferrous iron Oxidizing Bacteria (FeOB). Additionally, techniques such as hydrolysis of Fe^{3+} in the presence of the complexing agent urea are also employed by some researchers, though being less common. The following sections will provide a detailed discussion of these synthesis methods.

3.2 Rapid oxidation method

The rapid oxidation of ferrous sulphate using H_2O_2 (~30%–33%) coupled with the simultaneous hydrolysis of Fe^{3+} produces schwertmannite. This method is the most widely used due to its simplicity and efficiency, allowing for the rapid formation of schwertmannite without requiring strict experimental precautions. It is also the easiest technique for large-scale laboratory production, capable of generating kilogram quantities. Initially proposed by Regenspurg et al. (2004), this method has since been adopted by numerous studies (Burton et al., 2008; Davidson et al., 2008; Paikaray et al., 2011; Burton and Johnston, 2012; Schoepfer et al., 2017, 2019; Fan et al., 2019a, 2019b; Choppala and Burton, 2018; Vithana et al., 2018; Wang et al., 2020). It is particularly convenient for producing larger quantities of schwertmannite by simply increasing the initial concentration of ferrous sulphate.

Both O_2 and H_2O_2 act as oxidants for Fe^{2+} oxidation under aerobic conditions. The oxidation of Fe^{2+} by atmospheric O_2 in acidic environments, without the presence of iron-oxidizing bacteria, is an extremely slow process (Stumm and Morgan, 1995). Due to its higher oxidation potential (1.77 V for H_2O_2 compared to 0.40 V for O_2), H_2O_2 facilitates spontaneous schwertmannite formation (Rao et al., 1995), making it an excellent oxidant (As et al., 2024) (Eq. 3.1).

$$Fe^{2+} + 0.5H_2O_2 + H^+ \rightarrow Fe^{3+} + H_2O \qquad \text{(Eq. 3.1)}$$

This oxidation reaction is thermodynamically favourable in oxygen-saturated environments. The oxidation of Fe^{2+} to Fe^{3+} is an acid-consuming process (Eqs. 3.1 and 3.2), whereas the subsequent hydrolysis and ferric precipitate formation is an acid-generating process (Kirby and Brady, 1998). In total, the combined oxidation and hydrolysis reaction leads to an acidification of the suspension (Eq. 3.2)

$$8Fe^{2+} + 4H_2O_2 + xSO_4^{2-} + (8-2x)H_2O \rightarrow Fe_8O_8(OH)_{8-2x}(SO_4)_x + (16-2x)H^+ \qquad \text{(Eq. 3.2)}$$

where x denotes sulphate content.

Studies on schwertmannite synthesis using the H_2O_2 technique have employed varying concentrations of iron and sulphur, while maintaining the Fe/S ratio within the recommended range of 4.6–8.0. The specific ferrous iron sources and their concentrations used in different studies are summarized in Table 3.1. Upon the addition of H_2O_2, a red-orange or

TABLE 3.1 Variable iron and sulphate salts used for preparing schwertmannite under rapid oxidation method

Iron sources	*Strength*	*Volume of Milli-Q water*	*Volume of oxidant (H_2O_2/ $KMnO_2$)*	*H_2O_2 strength (wt%)*	*pH*	*References*
$FeSO_4{\cdot}7H_2O$	0.1 M	100 mL	0.75 mL	35	2.5	Barham (1997)
$FeSO_4$	0.065 M	500 mL	5 mL	35	4.2	Ooi et al. (2017)
$FeSO_4{\cdot}7H_2O$	0.059 M	1000 mL	5.3 mL	30	2.5	Song et al. (2015)
$FeSO_4{\cdot}7H_2O$	0.060 M	30 L	150 mL	30	2.4	Schoepfer et al. (2017)
$FeSO_4{\cdot}7H_2O$	0.108 M	50 L	800 mL	30	2.5	Johnston et al. (2016)
$(NH_4)_2Fe(SO_4)_2{\cdot}6H_2O$	0.144 M	1000 mL	1000 mL	33	3.0	Davidson et al. (2008)
$FeSO_4{\cdot}7H_2O$ + K_2SO_4	0.050 M + 0.008 M	150 mL	100 mL	1.5	2.5	Huang and Zhou (2012)
$FeSO_4{\cdot}7H_2O$	0.033 M	500 mL	2.65 mL	30	3.0	Zhang et al. (2016)
$FeSO_4{\cdot}7H_2O$	0.16 M	20 L	240 mL	32	2.5	As et al. (2024)

red-brown amorphous precipitate begins to form immediately, which turns dark red after 24 hours (Song et al., 2015). To produce larger quantities of schwertmannite, higher volumes of ferrous sulphate and H_2O_2 are utilized, with a slower H_2O_2 addition rate and continuous stirring to ensure thorough mixing (e.g., Johnston et al., 2016; Schoepfer et al., 2017). After reaching equilibrium at 24 hours, the schwertmannite precipitate is separated through centrifugation, followed by repeated washing with Milli-Q water to remove soluble ions, and then dried at 40°C.

Davidson et al. (2008) introduced a variation in schwertmannite synthesis at room temperature by using ferrous ammonium sulphate instead of ferrous sulphate, while the oxidant H_2O_2 remained unchanged. Similarly, the use of K_2SO_4 as a sulphate source, in addition to $FeSO_4 \cdot 7H_2O$, appears to have little effect on schwertmannite formation (Huang and Zhou, 2012). However, the rate of H_2O_2 addition, as demonstrated in several studies (e.g., Huang and Zhou, 2012; Liu et al., 2015), can significantly influence the type of product formed. Slow H_2O_2 addition at a rate of 0.2 mL min^{-1} resulted in K^+-jarosite formation, whereas spontaneous H_2O_2 addition yielded schwertmannite from the same $FeSO_4 \cdot 7H_2O$ and K_2SO_4 solution (Huang and Zhou, 2012). This finding was consistent with Liu et al. (2015), who observed that an Fe/S ratio in the resulting schwertmannite of 4.67 was achieved with 1.8 mL of H_2O_2 added over 2 hours, increasing to 5.04 with 0.36 mL added over 50 hours.

Schwertmannite has also been synthesized using water from the Smolnik sulphide mine in Slovakia (Mačingová and Luptáková, 2014), where the mine water was neutralized to pH 3.75 using 5 M NaOH before oxidation. To produce trace-metal-substituted schwertmannite (e.g., As), the same synthesis method was employed with an additional step of dissolving the desired amount of an arsenic-containing salt in the ferrous sulphate solution before adding H_2O_2. For instance, Johnston et al. (2016) dissolved 9 g of $Na_2HAsO_4 \cdot 7H_2O$ together with 1.5 kg of $FeSO_4 \cdot 7H_2O$ in 50 L of deionized water, yielding ~2500 mg kg^{-1} As(V)-schwertmannite upon H_2O_2 oxidation. Similarly, Regenspurg and Peiffer (2002) used 28.1 g of $Na_2HAsO_4 \cdot 7H_2O$ with $FeSO_4 \cdot 7H_2O$ to obtain a 20/1 molar ratio of $FeSO_4/AsO_4^{3-}$ in the final schwertmannite. Other studies have followed comparable synthesis procedures (e.g., Zhang et al., 2016; Wang et al., 2020).

In addition to H_2O_2, potassium permanganate ($KMnO_4$) has been used as an oxidant in recent studies due to its ease of handling and field-scale applicability (Cao et al., 2021). This method is similar to the H_2O_2 method, with $KMnO_4$ replacing H_2O_2 as the oxidant. Zhu et al. (2020) utilized both $KMnO_4$ and MnO_2 to oxidize ferrous solutions ($FeSO_4 \cdot 7H_2O$) containing fluoride (100–1500 mg L^{-1}), with an Fe^{2+}/F^- ratio of 2/1 to 6/1, and $KMnO_4$ or MnO_2/Fe(II) ratios of 0.1/1 to 1/1, at temperatures ranging from 20°C to 80°C.

The oxidation potential of $KMnO_4$ is significantly higher than that of H_2O_2, leading to higher yields of schwertmannite. In fact, less $KMnO_4$ is required compared to H_2O_2 to achieve a similar quantity of product, with a greater proportion of Fe^{2+} oxidized. However, a potential drawback of the $KMnO_4$ method is the possibility of Mn-doping at higher $KMnO_4$ concentrations, which may either inhibit schwertmannite formation or reduce its sorptive capacity. Schwertmannite produced via $KMnO_4$ oxidation is typically of high purity but exhibits poor crystallinity, which decreases further with increasing $KMnO_4$ concentrations. Both H_2O_2 and $KMnO_4$ concentrations affect particle size, with larger particles formed in the $KMnO_4$ method. For example, Cao et al. (2021) found that particle size increased from 1437 to 2348 nm as the H_2O_2/Fe^{2+} molar ratio increased from 0.124 to 0.992, compared to an increase from 865 to 2437 nm when the $KMnO_4/Fe^{2+}$ ratio increased from 0.0315 to 0.1585.

Procedure for schwertmannite synthesis using H_2O_2

1. Dissolve 10 g of $FeSO_4$ in 1 L of ultrapure Milli-Q water
2. Add 5 mL of H_2O_2 (~32%)
3. Perform reagent mixing under continuous stirring using a magnetic stirrer
4. Continue stirring for several hours at room temperature to allow for schwertmannite precipitation and to reach equilibrium, with the final pH of the slurry stabilizing between 2.5 and 2.8
5. Immediately after H_2O_2 addition, the schwertmannite precipitate will appear reddish orange, gradually turning darker red as mixing continues
6. Decant the supernatant and wash the newly precipitated schwertmannite four to five times with ultrapure Milli-Q water using centrifugation to remove excess iron and sulphur
7. Dry the cleaned precipitate at 30°C–40°C and store in polyethylene (PE) bottles until use

3.3 Dialysis method

The dialysis technique is the earliest method to synthesize schwertmannite in the laboratory. This method involves the hydrolysis of ferric ions in the presence of sulphate, followed by dialysis of the resulting precipitate using cellulose membrane dialysis tubes in Milli-Q water. Brady et al. (1986) used 0.02 M $Fe(NO_3)_3$ with 0, 250, 500, 1000, 1500, and 2000 mg L^{-1} sulphate (as Na_2SO_4) and hydrolyzed the solution for 12 minutes at 60°C, yielding SO_4/Fe solution ratios between 0 and 1.8. The precipitates were subsequently dialyzed in (1) solutions containing the same sulphate concentrations used during hydrolysis (0–2000 mg L^{-1}), and (2) Milli-Q water (DW), with the dialyzing medium being changed daily. The initial pH of the dialyzing solutions was adjusted to 6 to approximate natural stream conditions. Initial subsamples collected after 36 hours were identified as natrojarosite (Na-jarosite), while final precipitates harvested after 30 days of dialysis were identified as schwertmannite (named as amorphous ferric hydroxysulphate).

Bigham et al. (1990) later adopted this method by hydrolyzing 0.02 M $FeCl_3$ at nine different sulphate concentrations (0, 62, 125, 250, 500, 750, 1000, 1500, and 2000 mg L^{-1}) as Na_2SO_4 for 12 minutes at 60°C. After cooling to room temperature, the slurry was dialyzed for 30 days in deionized water using cellulose membranes with a pore radius of 2.4 nm. This approach aimed to determine the optimal sulphate concentration for schwertmannite formation. Schwertmann and Carlson (2005) followed a similar technique by dissolving 10.8 g of $FeCl_3 \cdot 6H_2O$ and 3 g of Na_2SO_4 in 2 L of Milli-Q water preheated to 60°C, allowing 12 minutes for equilibration. The deionized water was replaced daily during the 30-day dialysis period until the electric conductivity (EC) dropped below 5 µS cm^{-1}. This technique continues to be widely used in modern studies (Regenspurg et al., 2004; Boily et al., 2010; Fernandez-Martinez et al., 2010; Paikaray et al., 2011; Antelo et al., 2013; Dong et al., 2015; Li et al., 2018; Ying et al., 2020).

While the above procedure represents the traditional method, some studies have deviated slightly. For example, a shorter dialysis period of 7 days was used by Wang et al. (2015), Ooi et al. (2017), and Ying et al. (2020), while Dong et al. (2015) reduced the dialysis time to 3 days. These studies maintained other parameters constant, using 10.8 or 5.4 g of $FeCl_3 \cdot 6H_2O$ and 3.0 or 1.5 g of Na_2SO_4 in 2 L of deionized water. Additionally, some studies used ferric nitrate instead of ferric chloride, such as 0.04 M Na_2SO_4 and 0.04 M $Fe(NO_3)_3 \cdot 9H_2O$ by Khamphila et al. (2017) and 0.02 M Na_2SO_4 and 0.02 M $Fe(NO_3)_3 \cdot 9H_2O$ by Ooi et al. (2017), with dialysis periods of 30 and 7 days, respectively, after cooling the 60°C mixture to room temperature. Bigham et al. (1996) used a similar approach with 2 L of 0.02 M K_2SO_4 at 60°C and allowed 12 minutes before adding 0.02 M Fe^{3+} as $Fe(NO_3)_3 \cdot 9H_2O$.

Hydrolysis without dialysis was also performed by Yu et al. (2002) using Na_2SO_4 and $FeCl_3 \cdot 6H_2O/Fe(NO_3)_3 \cdot 9H_2O$ at various Fe/SO_4 ratios. Mazzetti and Thistlethwaite (2002) introduced a modification to the method by further heating the mixture for 10 minutes after schwertmannite formation. Dialysis temperature (25°C, 40°C, 50°C, 70°C, and 80°C), time (1, 3, 7, 10, and 15 days), and the presence of foreign ions (K^+ and NH_4^+) were evaluated by Ying et al. (2020). Schwertmannite was also synthesized in the presence of 0.1, 1.0, and 2.0 g L^{-1} K_2CrO_4 along with 5.4 g L^{-1} $FeCl_3 \cdot 6H_2O$ and 1.5 g L^{-1} Na_2SO_4 (Wan et al., 2018). This mixture, prepared in Milli-Q water at 60°C for 12 minutes and dialyzed for

30 days, yielded schwertmannite containing 0.22, 0.37, and 0.60 wt% Cr(VI) without altering its structure. This structural stability may be due to the replacement of SO_4^{2-} by Cr(VI), as observed by Regenspurg and Peiffer (2005) at 8.62 mM CrO_4^{2-}. Schwertmannite with an As/(As + S) ratio of 0.0–1.0 has also been synthesized using AsO_4/SO_4 ratios of 0/1000, 31/969, 62/938, 125/875, 250/750, 500/500, 750/250, and 1000/0. High concentrations of As(V) in solution hinder schwertmannite formation, as noted in several studies. Instead, ferric arsenate forms at high As(V)/Fe(III) molar ratios, particularly when As(V)/Fe(III) > 0.25 (Carlson et al., 2002).

Procedure for schwertmannite synthesis using dialysis

1. Heat 1 L of deionized water to 60°C in a water or oil bath
2. Add 5.4 g of $FeCl_3 \cdot 6H_2O$ and 1.5 g of Na_2SO_4 immediately, stirring continuously
3. Maintain the slurry temperature at 60°C for 12 minutes under constant stirring to ensure complete hydrolysis of Fe^{3+}
4. Cool the slurry to room temperature (22°C–25°C)
5. Transfer the suspension into dialysis tubes and dialyze for approximately 30 days in deionized water, using a suitable water tub
6. Replace the deionized water in the dialysis tub daily until the final conductivity reaches less than 20 $\mu S\ cm^{-1}$
7. Collect the precipitates after thoroughly washing with deionized water several times via centrifugation to remove excess salts
8. Dry the cleaned precipitate at 30°C–40°C and store it in PE bottles until use

---

3.4 Bacterial oxidation method

Bacterial oxidation of Fe^{2+} in a sulphate-rich medium is another widely used method for synthesizing schwertmannite. Typically, bacterial cultures are grown in suitable growth media with the necessary nutrients before being exposed to dissolved ferrous iron. The oxidation of Fe^{2+} to Fe^{3+} and the simultaneous hydrolysis of ferric ions result in a high yield of schwertmannite. During this process, the oxidation of Fe^{2+} consumes protons, raising the solution pH, while the formation of schwertmannite releases protons, causing a rapid decrease in pH. These opposing reactions help to maintain a relatively stable, yet acidic, pH throughout the synthesis. This laboratory synthesis technique mimics natural processes occurring in microbially active acidic mine drainages. In fact, chemical oxidation rates are 10^5–10^6 times slower at these pH values (Kawano and Tomita, 2001) than those observed in bacterial oxidation processes.

In the following, several bacterial oxidation synthesis methods are discussed to highlight the similarities and differences in the approaches.

Lazaroff et al. (1982) pioneered a method of synthesizing a precipitate from 0.2 M $FeSO_4$ through bacterial oxidation using *T. ferrooxidans*, which they referred to as "amorphous ferric hydroxysulfate". This precipitate displayed Fourier transform infrared spectroscopy bands at 3270–3340 cm^{-1} (OH stretching), 1619–1630 cm^{-1} (H_2O deformation), 1110–1120 cm^{-1} ($\nu_3(SO_4)$), 970–978 cm^{-1} ($\nu_1(SO_4)$), and 645–650 cm^{-1} ($\nu_4(SO_4)$), which closely matched the infrared bands of schwertmannite. The bacterial oxidation was performed using 100 mL of 0.1–0.2 M $FeSO_4$ at pH 2.5. Flasks containing 10^{11} cells of *T. ferrooxidans* suspended in H_2SO_4 (pH 2.5) were incubated for 10 days at 23°C with vigorous shaking. Later, Bigham et al. (1990) applied a similar approach incubating *T. ferrooxidans* in 9K medium to oxidize $FeSO_4{\cdot}7H_2O$ at pH 2–3. The 9K medium contains a variety of salts among them also the jarosite-forming cations K^+, Na^+, and NH_4^+. Acclimated bacterial cells were added to a 4% $FeSO_4$ solution at room temperature, at an initial pH of 2.9 and a redox potential of 270 mV. After 10 days, NH_4-jarosite formed as the pH dropped to 1.4, and the redox potential stabilized at 825 mV. The brown colloidal suspension was flocculated by titration to pH 4 with 0.1 M NaOH, resulting in the formation of schwertmannite.

Subsequent studies employed *A. ferrooxidans*, another Fe^{2+}-oxidizing bacterium frequently found in acid mine drainage (AMD) environments (Benner et al., 2000; Fortin and Langley, 2005). Like *T. ferrooxidans*, *A. ferrooxidans* is an acidophilic, chemolithotrophic bacterium that derives energy from the oxidation of Fe^{2+} to Fe^{3+}. Optimal growth occurs at pH 2.0–2.5, although the bacterium can survive within a pH range of 1.5–6.0.

The composition of the 9K medium, which is commonly used for growing also *A. ferrooxidans*, was slightly varied in several studies, as summarized in Table 3.2. Before inoculating the medium with *A. ferrooxidans*, the 9K medium is adjusted to a pH of 2.0–2.5 using an appropriate acid, such as H_2SO_4. The $FeSO_4$ solution is filter-sterilized and added to the autoclaved medium. The bacterial culture is then incubated for a specified period, ranging from a few hours (48 hours, Huang and Zhou, 2012) to several days (5 days, Zhou et al., 2012), at room temperature using rotary shakers. After the incubation period, the precipitates are removed by filtration, and the bacterial cells are separated from the filtrate through centrifugation.

Wang et al. (2006) used *A. ferrooxidans* to produce schwertmannite at NH_4^+ concentrations of 5.4 and 11.4 mM $(NH_4)H_2PO_4$ in 4.05 mM H_2SO_4. After 7 days of incubation, the schwertmannite that precipitated at an NH_4^+ concentration of 5.4 mM was reddish in colour, while the precipitate obtained in the presence of 11.4 mM NH_4^+ was yellowish. The specific surface areas (SSA) of the samples were 12.8 and 28.5 $m^2\ g^{-1}$,

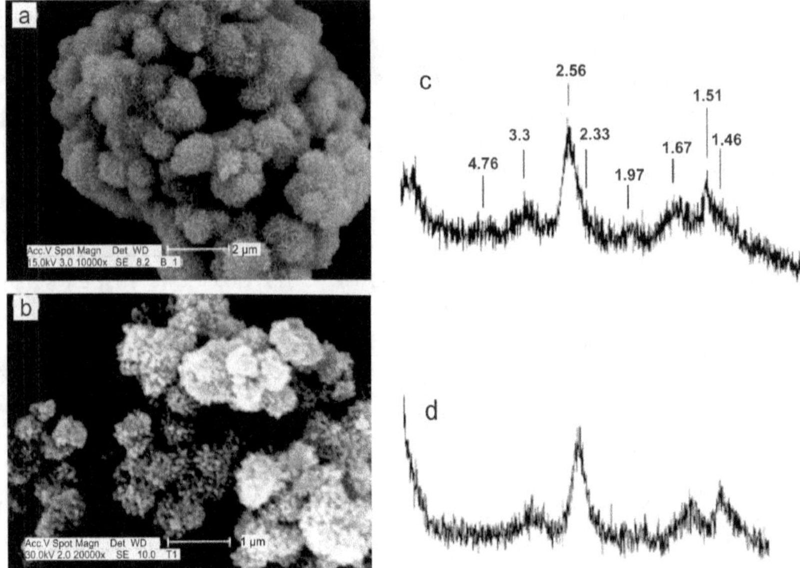

FIGURE 3.1 Scanning electron micrographs of schwertmannite precipitated in *A. ferrooxidans* culture media (a) at 36°C after 7 days in the presence of 5.4 mM NH_4^+, and (b) at 36°C after 19 days in the presence of 11.4 mM NH_4^+ with their corresponding X-ray diffractograms (c and d, respectively) (ref. Wang et al., 2006).

respectively. The X-ray diffraction patterns showed stronger intensities for the eight diffraction peaks in the sample from the lower NH_4^+ concentration (Figure 3.1). A similar synthesis method was followed by Pállová et al. (2010) at a pH of ~2.8, and by Gramp et al. (2009), who used 6.1 and 11.4 mM $NH_4H_2PO_4$, respectively, while keeping the other ionic concentrations the same as in Wang et al. (2006). Liao et al. (2011) used *A. ferrooxidans* LX5 at a cell density of ~2.5×10^8 cells mL^{-1}, producing schwertmannite within 1 hour of incubation from 0.144 M $FeSO_4$ at 28°C.

Schwertmannite was the sole mineral at pH 3, while traces of jarosite appeared at lower pH levels (pH 2.3, 2.6) and of goethite at higher pH levels (pH 3.3, 3.6) (Bigham et al., 1996). Bacterial oxidation of Fe^{2+} was indicated by an increase in Eh from ~500 mV to ~580 mV. Similarly, variable Fe^{2+} concentrations of 20, 40, 80, and 160 mM (as $FeSO_4 \cdot 7H_2O$) were tested using *A. ferrooxidans* at a cell density of 6×10^7 cells mL^{-1} at pH ~ 2.5 (Bai et al., 2012). While this yielded pure schwertmannite, the addition of monovalent cations (K^+, Na^+, and NH_4^+) shifted the formation towards jarosite. However, the 3 g of $(NH_4)_2SO_4$ in the 9K medium (cf. Table 3.2) was insufficient to produce NH_4-jarosite, though it was essential for bacterial metabolism. Similar observations were made by Gramp et al. (2009) in the presence of 1.3 g L^{-1} $NH_4H_2PO_4$ in the 9K medium.

Bai et al. (2012) studied the dependence of schwertmannite formation on Fe^{2+} concentrations as well as the ratio between Fe^{2+} concentration and monovalent cations. Schwertmannite was the only mineral phase formed at an initial Fe^{2+} concentration of 160 mM in the presence of up to 320 mM Na^+, 120 mM NH_4^+, and 6.4 mM K^+. Also at lower Fe^{2+} concentrations, schwertmannite formed, even at high K^+ levels. The presence of K^+ promotes K-jarosite formation 200 times greater than Na-jarosite and 75 times faster than NH_4-jarosite. Based on the Fe/M^+ ratio, Bai et al. (2012) proposed a stoichiometric relationship to estimate the critical Fe/M^+ ratio that determines the threshold for schwertmannite formation. The proposed stoichiometric equations for K^+ (Eq. 3.3) and NH_4^+ (Eq. 3.4) are as follows:

$$Y = -0.8077 - 0.04257x + 0.00617x^2 \qquad \text{(Eq. 3.3)}$$

$$Y = 0.0354 - 0.00295x - 7.407E - 5x^2 \qquad \text{(Eq. 3.4)}$$

where Y = Fe/NH_4^+ or Fe/K^+ molar ratio and x = initial $Fe(II)_{aq}$ concentrations (mM).

When the ratios Fe/NH_4^+ or Fe/K^+ are larger than Y, schwertmannite was predominant; while for Fe/NH_4^+ or Fe/K^+ less than Y, mixtures between schwertmannite and jarosite formed.

In addition to the increase in Eh (Bigham et al., 1990, 1996), a decrease in aqueous Fe^{2+} concentrations is indicative of its oxidation over time. Faster oxidation at pH > 2 was observed by Liao et al. (2009), with 100% oxidation achieved within 36 hours at a rate of 3.9 mmol L^{-1} h^{-1}, compared to 1.2 mmol L^{-1} h^{-1} at pH ≤ 1.8, where complete oxidation took 96 hours. In addition to Fe^{2+} oxidation, researchers found that when the initial pH was ≥2.81, the pH decreased continuously to a constant value of ~1.7. In contrast, when the initial pH was ≤2.58, there was an initial rise in pH followed by a drop to ~1.7. This behaviour was attributed to the kinetic decoupling of Fe^{2+} oxidation to Fe^{3+}, which is an acid-consuming process, and the hydrolysis of Fe^{3+} to ferric precipitates, which releases protons and lowers the pH. The pH drop and decrease in Fe^{2+} concentration at pH ≥ 2.81 indicate faster Fe^{3+} precipitation compared to Fe^{2+} oxidation, while at pH ≤ 2.58, Fe^{2+} oxidation preceded faster than Fe^{3+} precipitation.

Different bacterial communities affect the nature of the products in different ways. Maillot et al. (2013) produced an As(V)–Fe(III) oxyhydroxysulphate phase from Carnoulès mine drainage, France, using *Thiomonas* sp. Tm B2 and Tm B3 strains, consistent with earlier work by Morin

TABLE 3.2 Concentration of salts used by different researchers to prepare the 9K medium for cultivating bacterial cells able to mediate microbial oxidation of Fe^{2+} (as $FeSO_4{\cdot}nH_2O$) under acidic conditions resulting in the formation of schwertmannite or jarosite

Bacterial species	*$FeSO_4{\cdot}7H_2O$*	*$(NH_4)_2SO_4$*	*KCl*	*K_2HPO_4*	*$MgSO_4{\cdot}7H_2O$*	*$Ca(NO_3)_2$*	*pH*	*References*
A. ferrooxidans	*44.3 g L^{-1}*	*3 g L^{-1}*	*0.1 g L^{-1}*	*0.5 g L^{-1}*	*0.5 g L^{-1}*	*0.01 g L^{-1}*	~2.5	Bai et al. (2012)
A. ferrooxidans	44.2 g L^{-1}	3.5 g L^{-1}	0.119 g L^{-1}	0.058 g L^{-1}	0.583 g L^{-1}	0.0168 g L^{-1}¶	~2.5	Qiao et al. (2017)
A. ferrooxidans	21.87 g L^{-1}*	3 g L^{-1}	0.1 g L^{-1}	0.5 g L^{-1}	0.5 g L^{-1}	0.01 g L^{-1}	2.5	Zhou et al. (2012)
A. ferrooxidans	44.2 g L^{-1}	3.5 g L^{-1}	0.119 g L^{-1}	0.058 g L^{-1}	0.583 g L^{-1}	0.017 g L^{-1}¶	2.5	Huang and Zhou (2012)
A. ferrooxidans	33.36 g L^{-1}	1.3 g L^{-1}Π	–	–	0.394 g L^{-1}	–	~2.0–2.5	Gramp et al. (2009)
T. ferrooxidans	40 g L^{-1}*	2 g L^{-1}	0.1 g L^{-1}	0.25 g L^{-1}Δ	0.25 g L^{-1}₸	0.01 g L^{-1}	2.9	Bigham et al. (1990)
A. ferrooxidans	33.36 g L^{-1}	0.62/1.3 g L^{-1}Π	–	–	0.394 g L^{-1}	–	~2.5	Wang et al. (2006)
A. ferrooxidans	33.36 g L^{-1}	0.81 g L^{-1}Π	–	–	0.394 g L^{-1}	–	~2.8	Pállová et al. (2010)
T. ferrooxidans	27.80 g L^{-1}	0.396 g L^{-1}	–	0.22 g L^{-1}	0.394 g L^{-1}	–	~2.0	Bigham et al. (1996)
T. ferrooxidans	15.19–30.38 g L^{-1}*	4.29 g L^{-1}	0.143 g L^{-1}	0.712 g L^{-1}	0.712 g L^{-1}	0.014 g L^{-1}	2.5	Lazaroff et al. (1982)
A. ferrooxidans	44.7 g L^{-1}	3.0 g L^{-1}	0.1 g L^{-1}	0.5 g L^{-1}	0.5 g L^{-1}	0.01 g L^{-1}	2.0	Zhu et al. (2020)

¶ as $Ca(NO_3)_2{\cdot}4H_2O$; * as $FeSO_4$; Π as $NH_4H_2PO_4$; Δ as K_2SO_4; ₸ as $MgSO_4$,—represents no information available for these salts.

et al. (2003). The strains completely oxidized As(III) to As(V) over 7 days, achieving solid-phase As/Fe ratios of 0.63 and 0.8, respectively, without detectable As(III). In contrast, in the presence of the strain *A. ferrooxidans* Af CC, As(III)-rich schwertmannite formed with an As/Fe ratio of 0.19 and no detectable As(V), suggesting that this strain does not oxidize As(III) to As(V) but promotes Fe^{2+} oxidation to Fe^{3+}.

Overall, the microbial-mediated oxidation of Fe^{2+} and subsequent formation of schwertmannite follows a similar mechanism across different procedures. Variations occur in the composition of the 9K medium, the inoculum volume (e.g., 5% in Huang and Zhou, 2012; Zhou et al., 2012; Liu et al., 2015, vs. 10% in Bai et al., 2012), cell density of *A. ferrooxidans* (e.g., 2.5×10^8 mL^{-1} in Liao et al., 2011, vs. 6×10^7 mL^{-1} in Bai et al., 2012, or ~1×10^7 mL^{-1} in Liu et al., 2015, vs. ~3×10^7 mL^{-1} in Qiao et al., 2017, or 1×10^8 mL^{-1} in Gan et al., 2015), and duration of incubation (e.g., 1 hour in Liao et al., 2011, vs. 48 hours in Huang and Zhou, 2012, 60 hours in Qiao et al., 2017, 5 days in Zhou et al., 2012, and 7 days in Gramp et al., 2009). Parameters such as the drying temperature of the precipitate (40°C–60°C), the number of washing steps (three to five times), and the centrifugation time (10–40 minutes) typically do not affect the nature of the products. However, *A. ferrooxidans* in 9K medium does not always produce schwertmannite at a pH of ~2.0, as shown by Ma and Lin (2012), where the presence of metals such as Pb, Mn, Cd, As, P, Mg, Cu, and Zn may have inhibited schwertmannite formation.

During the microbial oxidation of Fe^{2+}, its concentration decreases sharply within hours, with complete removal of Fe^{2+} occurring over a few days (Ma and Lin, 2012). Despite the rapid oxidation of Fe^{2+} to Fe^{3+}, the total iron concentration in solution remains high because Fe^{3+} does not precipitate immediately in acidic conditions, especially at $pH < 2$, in the absence of cations that favour jarosite formation. Dissolved O_2 levels also decrease rapidly as bacterial metabolism consumes oxygen.

Biosynthesis experiments performed in the presence of $AlPO_4$ at varying ratios between $FeSO_4 \cdot 7H_2O$ and $AlPO_4$ (20/0.1, 20/1.0, 20/2.0, and 20/5) demonstrated the occurrence of an $AlPO_4$-modified schwertmannite (Gan et al., 2015). Additionally, anionic substitution was performed to produce schwertmannite at Cl^-/SO_4^{2-} ratios of 0, 3, 6, and 10, using *A. ferrooxidans* and $FeCl_2$ and $FeSO_4$ salts in varying proportions (Xiong et al., 2020). The oxidation rate slowed as the Cl^- content increased, with only half of the Fe^{2+} oxidized after 36 hours in $Cl^-/SO_4^{2-} = 10$ reactors, while nearly complete oxidation was achieved in reactors without Cl^-. The pH dropped from 3.0 to ~1.6–1.9, consistent with schwertmannite formation, and the decrease was more pronounced in Cl^--containing reactors.

Procedures for schwertmannite synthesis using microbes

1. Prepare a feed solution with the desired composition listed in Table 3.2 for bacterial growth. Minimize the use of monovalent cations to prevent jarosite precipitation
2. Prepare 1 L of $FeSO_4·7H_2O$ solution at the desired concentration in a glass reaction vessel
3. Add 10 mL of the feed solution containing acclimated bacterial cells
4. Adjust the pH to ~2.0 with concentrated H_2SO_4, ensuring minimal effect on total solution volume
5. Stir the mixture continuously at room temperature until brown precipitates begin to form
6. Maintain the system pH between ~2.0 and 3.0 using NaOH or H_2SO_4, possibly with an auto-titrator
7. Continuously monitor the redox potential (Eh) and dissolved Fe^{2+} concentration to track oxidation progress. Eh should increase, and Fe^{2+} concentration should decrease over time, indicating Fe^{2+} oxidation
8. Collect the ochreous precipitates, wash them by centrifugation, and dry at 40°C–60°C for further study

3.5 Urea hydrolysis method

Schwertmannite synthesis has also been performed through hydrolysis of ferric ions under laboratory conditions using urea as a reactant to raise the pH. The advantage of using urea is that the reaction products are pure and uniform with respect to their shape (Šubrt et al., 2011). This synthesis pathway is applied under atmospheric conditions by use of ferric sulphate salts and has been found to be effective in yielding large quantities of schwertmannite. In this method, an acidic iron salt solution of the desired concentration is preheated to ~70°C and titrated with a urea solution to facilitate complete hydrolysis of iron into schwertmannite. The specific details of iron salts, urea concentrations, and reaction times used in different studies are summarized in Table 3.3.

Generally, the preheated ferric solution is titrated slowly with the urea solution over 2–4 hours, during which the colour of the slurry changes from red to brownish-yellow as precipitation begins (Eskandarpour et al., 2008; Ooi et al., 2017). This technique has also been applied to mine drainage samples from Malý Šobov, Slovakia, using a urea/Fe ratio of 50 (Šubrt et al., 2011). The success of this urea hydrolysis method strongly depends on the urea/Fe ratio, temperature, and reaction time. Schwertmannite, along with goethite, jarosite, and ferrihydrite, can form from the same $Fe_2(SO_4)_3$ + urea mixture, depending on the experimental conditions (Šubrt et al., 1999, 2011). Šubrt et al. (1999) investigated urea/Fe ratios

TABLE 3.3 Details of experimental parameters for schwertmannite precipitation under urea hydrolysis method

Iron sources	*Weight of iron salt*	*Volume of Milli-Q water*	*Strength of urea*	*Volume of urea solution*	*References*
$Fe_2(SO_4)_3{\cdot}5H_2O$	12.5 g	250 mL	5 M	250 mL	Dey et al. (2014)
$Fe_2(SO_4)_3{\cdot}5H_2O$	25 g	500 mL	2.5 M	500 mL	Eskandarpour et al. (2008)
$Fe_2(SO_4)_3$	6 g	150 mL	5 M	150 mL	Ooi et al. (2017)
$Fe_2(SO_4)_3{\cdot}5H_2O$	250 g	2000 mL	8.3 M	700 mL	Šubrt et al. (1999)

ranging from 1.7 to 27 and found that schwertmannite starts precipitating when the urea/Fe ratio is ≥3.4, at a pH of 3–6. Therefore, careful control of experimental variables is essential to ensure the exclusive formation of schwertmannite.

In a two-step procedure, the urea-based hydrolysis method was applied to embed cobalt ferrite nanoparticles into schwertmannite to enhance its magnetic properties for improved arsenic removal (Dey et al., 2014). To this, $CoFe_2O_4$ nanoparticles were synthesized via a co-precipitation from a mixture of $Co(NO_3)_2{\cdot}6H_2O$ and a $Fe(NO_3)_3{\cdot}9H_2O$ at 80°C. Schwertmannite was synthesized through homogeneous hydrolysis of $Fe_2(SO_4)_3{\cdot}5H_2O$, with urea as the neutralizing agent. The modified schwertmannite produced by this method lost its characteristic XRD peaks and exhibited a pH_{PZC} of 3.7, with a $CoFe_2O_4$/schwertmannite ratio of 5–6. The incorporation of cobalt ferrite enhanced the paramagnetic properties of the schwertmannite, with a saturation magnetization of 13.95 emu g^{-1} and a coercivity value of 969.55 G, while retaining the material's characteristic pin-cushion morphology.

A generalized synthesis approach for schwertmannite via urea hydrolysis is outlined below. This method is expected to yield schwertmannite with typical morphology, composition, and stoichiometry similar to other synthesis techniques:

Procedure for schwertmannite synthesis using urea
1. *Dissolve 25 g of $Fe_2(SO_4)_3{\cdot}5H_2O$ in 500 mL of distilled water using a 1 L glass beaker*
2. Heat the $Fe_2(SO_4)_3{\cdot}5H_2O$ solution to ~70°C on a temperature-controlled hot plate
3. Dissolve 150 g of urea in 500 mL of distilled water in a separate 1 L glass beaker

(*Continued*)

Procedure for schwertmannite synthesis using urea
4. Using a burette, add the urea solution dropwise into the $Fe_2(SO_4)_3{\cdot}5H_2O$ solution over 2–4 hours, ensuring thorough mixing by continuous stirring
5. Continue stirring the slurry for an additional 1 hour after completing the urea addition to achieve complete hydrolysis of $Fe_2(SO_4)_3{\cdot}5H_2O$
6. Wash the slurry with Milli-Q water five times, discarding the supernatant after each wash, to obtain a pure precipitate
7. Oven-dry the solid precipitate and store it in PE storage bottles

3.6 Industrial biosynthesis method

Attempts have been made to synthesize schwertmannite at a pilot plant scale using mine waters instead of Milli-Q water. These waters, enriched with Fe^{2+} (300–400 mg L^{-1}) and SO_4^{2-} (1700–2100 mg L^{-1}), were sourced from lignite mine effluents at the Lusatian mine, Nochten, Freiburg, SE Germany (Janneck et al., 2010). This approach was considered economical due to the absence of treatment chemicals and additional nutrient supply, as iron-oxidizing bacteria in the system utilized dissolved CO_2 as a carbon source.

The pilot plant was designed as an oxidation basin with volume of 8.14 m^3 and was equipped with removable growth carriers. Additionally, an aeration and precipitation tank was incorporated. A high-flow circulation pump (~30 m^3 h^{-1}) was used to aerate the water inflow intensively. The pH in the basin fluctuated between 2.3 and 2.7. A chain cleaner at the bottom of the oxidation basin continuously removed the formed schwertmannite, directing it into a sludge collection channel. From there, the sludge was pumped into a storage tank where it was thickened. The thickened sludge could then either be recycled back to the oxidation basin or transferred to large bags for dewatering by gravity.

The pilot plant was designed as a hybrid bioreactor, combining the characteristics of a fixed-bed reactor that prevented clogging with unrestricted circulation within the oxidation basin (Figure 3.2). This study recommended that if two-third of the iron load in the treatment plant (Lusatia lignite mine, Germany) could be precipitated as schwertmannite, a recovery of 8000–14,000 tonnes per annum of schwertmannite would be possible over a period of at least 30 years (Janneck et al., 2010).

The schwertmannites formed during these process exhibit identical properties as found in other synthesis methods. For example, Hedrich et al. (2011) reported Munsell colour index of 7.5 YR 5.5/8 with hedge-hog morphology of 350–700 nm diameter and surface whiskers of 260 nm length for schwertmannites formed in mine water treatment plants by using bacterial strain *Ferrovum myxofaciens* in Nochten, Lusatia, Germany (Figure 3.3a). The SAED electron diffractogram revealed a characteristic ring

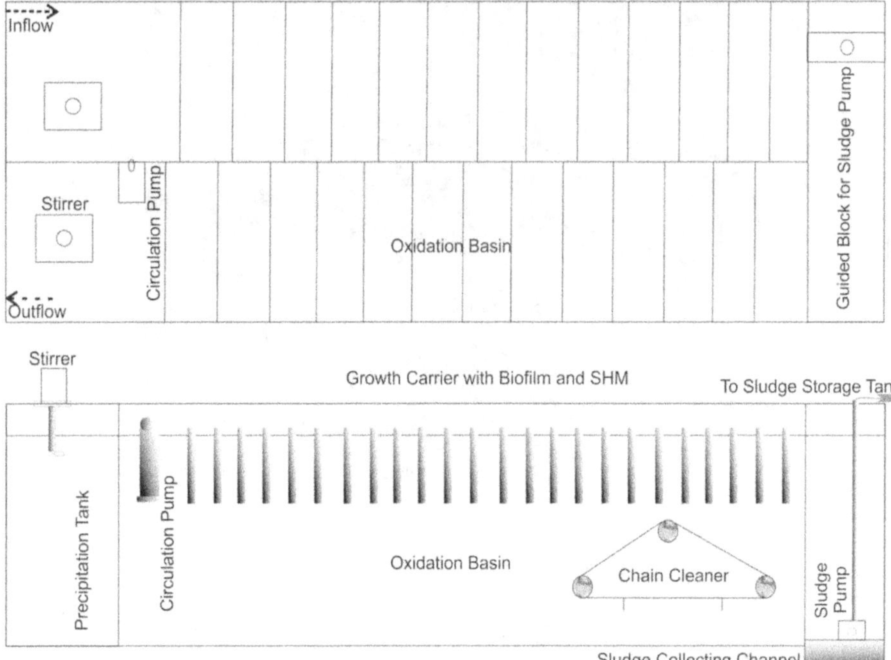

FIGURE 3.2 A schematic sketch showing the design of a pilot plant that produces schwertmannite in large scale through bacterial oxidation of high iron- and sulphate-rich mine effluents (ref. Janneck et al., 2010).

pattern with d-spacing of the first ring of 0.254 nm (Figure 3.3b). The predominant mineral formed in the pilot plant was schwertmannite at pH 3.0 with Fe/S ratio of 6.1–6.6, while jarosite and goethite sometimes occurred in trace amounts. The study proposed that K^+-jarosite formed in this process was the transformed product of schwertmannite.

3.7 Other synthesis methods

The five synthesis methods described earlier are commonly used for studying properties and characteristics of schwertmannite, particularly the first three techniques. However, additional synthesis methods are also practiced in various laboratories, differing significantly from traditional approaches. A few such methods are listed below.

Different approaches comprise the hydrolysis of ferric sulphate solutions at elevated temperature and pH ~1.7. Loan et al. (2004, 2005) prepared a solution of 250 mL containing 0.4 g of $Fe_2(SO_4)_3{\cdot}xH_{2O,}$ while Kumpulainen et al. (2008) used 500 mL containing 2.506 g of $Fe_2(SO_4)_3{\cdot}xH_2O$. These solutions were placed in Nalgene bottles (pH ~ 1.7) and subjected

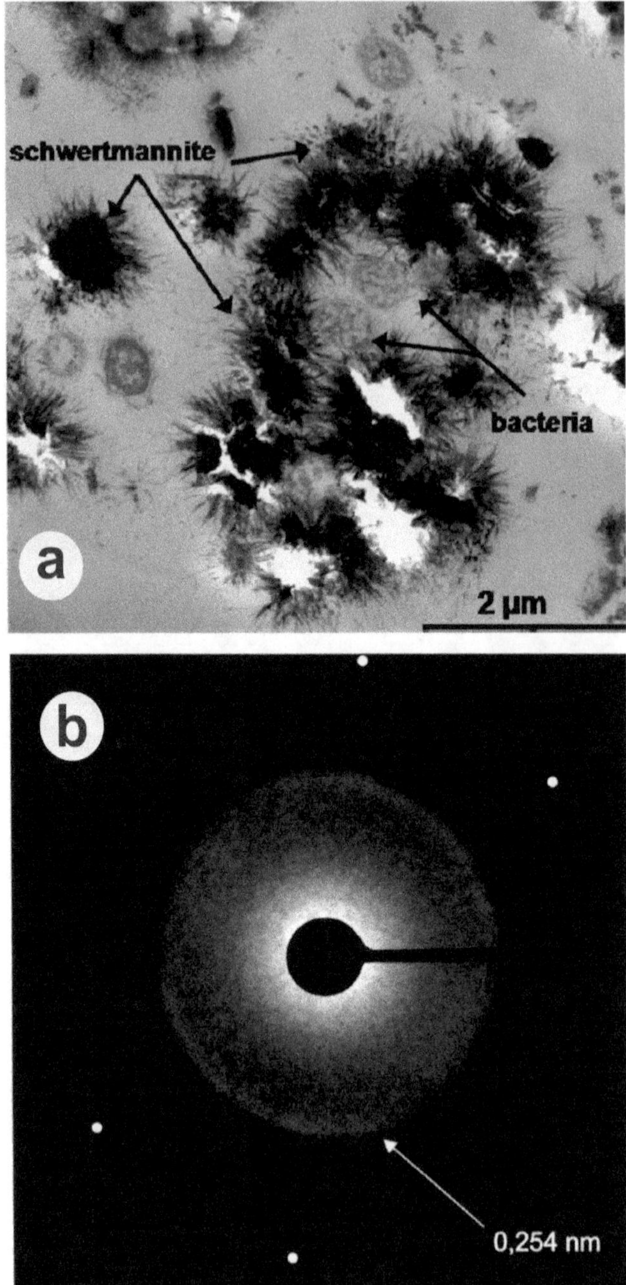

FIGURE 3.3 Transmission electron micrographs of ThO_2-stained thin sections of schwertmannite mineral aggregates surrounding Fe(II)-oxidizing bacteria (a) and SAED of schwertmannite microaggregates (b) (ref. Hedrich et al., 2011).

to a mechanical bottle-roller in a water bath at 85°C for 1 hour, resulting in schwertmannite precipitation. Loan et al. (2004) who used this method first stated that this method avoids the complications caused by chloride and nitrate solutions that favour akaganéite or a reported oxyhydroxynitrate phase. This technique was later applied by Maillot et al. (2013) and Cruz-Hernández et al. (2019) to produce As-containing schwertmannites in the presence of $NaAsO_2$ and Na_2HAsO_4 in addition to $Fe_2(SO_4)_3 \cdot xH_2O$.

Hockridge et al. (2009) used the same procedure and observed globular schwertmannite precipitates forming within 5 minutes at 85°C, with surface whiskers developing after 10 minutes and becoming predominant after 60 minutes (see Figure 2.3 in chapter 2). This observation suggests that the needle-like structures do not grow from the central nucleus but rather develop on the surface of large spherical aggregates, increasing in size over time from <50 nm at 10 minutes to 200 nm at 30 minutes (Figure 3.4; Hockridge et al., 2009).

Instead of dissolving $Fe_2(SO_4)_3$ in deionized water, Houngaloune et al. (2014, 2015, 2017) used 10 mM H_2SO_4 to dissolve the ferric salt at 65°C,

FIGURE 3.4 TEM images showing the development of schwertmannite needles after 1o minutes (a), 15 minutes (b), 30 minutes (c) and 60 minutes (d) reaction time. Image (d) shows a mixture of schwertmannite and goethite-like aggregates (ref. Hockridge et al., 2009).

adjusting the pH to 3–4 with 1 M Na_2CO_3 to promote schwertmannite precipitation. Similarly, Song et al. (2015) used preheated Milli-Q water to dissolve $Fe_2(SO_4)_3$ in a water bath at 85°C for 1 hour.

Cutting et al. (2009) employed a temperature ramping method to synthesize schwertmannite by heating a 10 mM $Fe_2(SO_4)_3{\cdot}9H_2O$ solution to 70°C at a rate of 35°C min^{-1}. The solution was maintained at 70°C for 8 minutes to promote the formation of ferric sulphate polymers. In a similar approach, Šubrt et al. (1999) added hexamethylene tetraamine dropwise over 30 minutes to 500 mL of 0.183 M ferric sulphate preheated to 80°C. The suspension was stirred for additional 8 hours before collecting the precipitates.

Tresintsi et al. (2012) synthesized schwertmannite at the kilogram scale using a two-stage continuous-flow reactor. The process involved oxidation, hydrolysis, and precipitation of ferrous salts ($FeSO_4{\cdot}H_2O$ and $FeCl_2{\cdot}4H_2O$) under redox conditions of 380–440 mV and a pH range of 3–5.5. The reactor produced approximately 0.25 kg of schwertmannite per hour at an inflow rate of 10 L h^{-1} of 40 g L^{-1} $FeSO_4{\cdot}H_2O$ solution. NaOH (30% w/w) was added to control the pH, while H_2O_2 (50% w/w) was used for redox control. This method produced schwertmannite with a spherical hollow structure, approximately 150 nm in diameter, within the reactor's pH range of 3.0–5.5.

Reichelt and Bertau (2015) employed a microjet mixing device, as shown in the schematic diagram below (Figure 3.5). This device features a microjet mixer with a nozzle diameter of 50–100 µm and two HPLC pumps, with compressed air fed into the mixer. Key variables studied included the ratio between NaOH and Fe^{3+} added (NaOH/Fe^{3+} ratio, 1.875–5.625), the age of the $Fe_2(SO_4)_3$ solution (0 to 7 days), the jet diameter (50–100 µm), the ferric sulphate concentration (0.016–0.8 M), and the fluid flow rate (10–30 mL min^{-1}). The study found that an NaOH/Fe^{3+} ratio of ~1.4–2.6

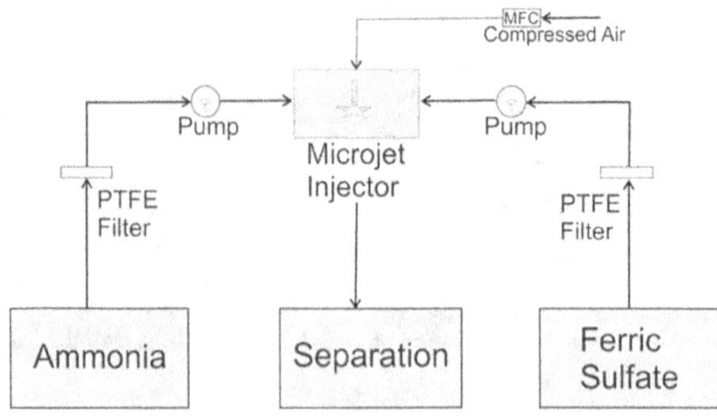

FIGURE 3.5 Schematic diagram of the microjet induced schwertmannite synthesis (modified after Reichelt and Bertau, 2015).

resulted in a pH range of 2.4–3.5, which was conducive to schwertmannite formation. Beyond this range, a sharp increase in pH and alkalinity led to the formation of ferrihydrite.

Ying et al. (2020) synthesized schwertmannite through the hydrolysis of Fe^{3+} using NaOH by rapidly mixing 2 mL of 0.8 M NaOH with an equal volume of 0.4 M $Fe_2(SO_4)_3$, with an Fe/S ratio of 0.67 and a pH of ~2.5. This mixture was hydrolyzed at 25°C for different durations of 3.2, 8.7, 45, 50, 59, 70, and 140 minutes to investigate the impact of hydrolysis time. Initially, ferrihydrite formed that quickly transformed into schwertmannite within 45 minutes. In contrast, the addition of ferrihydrite to the sulphate-rich medium without the hydrolysis step did not yield schwertmannite. The effect of the hydrolysis rate on schwertmannite formation was studied by adding 0.2 M NaOH at rates of 33.33, 6.67, and 3.33 µM min^{-1} to a mixed solution of $Fe(NO_3)_3$ and Na_2SO_4. Slower hydrolysis rates led to a higher crystallinity with larger aggregates, while faster rates resulted in poor crystallinity with blocky aggregates.

Schwertmannite generation also occurred accidentally, when the appropriate ionic concentrations and pH are achieved. For example, traces of schwertmannite were found while synthesizing goethite (α-FeOOH) from a 1.0 M $FeSO_4$ medium in the presence of Ni^{2+}, Cu^{2+}, and Cr^{3+}, with metal/Fe ratios of 0.0–0.1. At a Cr^{3+}/Fe^{3+} ratio ≥0.04, schwertmannite was observed, although α-FeOOH remained the dominant phase. This result was attributed to the similar ionic radii of Cr^{3+} (0.063 nm) and Fe^{3+} (0.064 nm) and the drop in pH from 7 to 4. A similar, but smaller effect was observed in the presence of Cu^{2+} (0.072 nm).

3.8 Summary

Various methods have been developed to synthesize schwertmannite under aerobic laboratory conditions, both abiotic and biotic. The majority of these methodologies focus on the oxidation of ferrous iron and simultaneous hydrolysis, using different oxidants such as H_2O_2 or iron-oxidizing bacteria in sulphate-rich aqueous media. Some methods also involve the hydrolysis of ferric iron, especially in dialysis-based techniques. Typically, the pH of the synthesizing medium is maintained between ~2.5 and 4.5, with dissolved sulphate available either naturally or as an added source, such as Na_2SO_4.

The dialysis method was one of the first proposed approaches for synthesizing schwertmannite, involving the use of cellulose membrane dialysis tubes over periods ranging from 3 to 30 days. In this method, ferric chloride and sodium sulphate solutions of the desired strength are mixed at 60°C for approximately 12 minutes, and then cooled to room temperature before being dialyzed in deionized water for up to 30 days. The deionized water is replaced daily until the conductivity reaches <20 μS cm^{-1}.

In the urea hydrolysis method, ferrous sulphate solutions are hydrolyzed by the dropwise addition of urea. The temperature is typically maintained at ~70°C, and reagent mixing is conducted for ~4 hours. In another method, the rapid oxidation of ferrous ions by H_2O_2, followed by the simultaneous hydrolysis of ferric ions, produces schwertmannite. The amount of schwertmannite synthesized by this method can be controlled by varying the concentration of ferrous sulphate, which is subsequently oxidized by the addition of 30%–33% H_2O_2. Microbial oxidation of ferrous ions by *A. ferrooxidans* and *T. ferrooxidans* has also been practiced to synthesize schwertmannite under ambient conditions. Besides these traditional synthesis methods, some studies have successfully developed alternative procedures for schwertmannite synthesis.

References

Antelo J, Fiol S, Gondar D, Pérez C, López R and Arce F (2013). Cu(II) incorporation to schwertmannite: Effect on stability and reactivity under AMD conditions. *Geochimica et Cosmochimica Acta* 119: 149–163.

As K, Peiffer S, Uhuegbue PO, Joshi P, Kappler A, Marouane B and Hockmann K (2024). Sulfate affinity controls phosphate sorption and the prototransformation of schwertmannite. *Chemical Geology* 653: 122043.

Bai S, Xu Z, Wang M, Liao M, Liang J, Zheng C and Zhou L (2012). Both initial concentrations of Fe(II) and monovalent cations jointly determine the formation of biogenic iron hydroxysulfate precipitates in acidic sulphate-rich environments. *Materials Science and Engineering* C 32: 2323–2329.

Barham BJ (1997). Schwertmannite: A unique mineral, contains a replaceable ligand, transforms to jarosites, hematites, and/or basic iron sulphate. *Journal of Material Research* 12: 2751–2757.

Benner SG, Gould WD and Blowes DW (2000). Microbial populations associated with the generation and treatment of acid mine drainage. *Chemical Geology* 169: 435–448.

Bigham JM, Carlson L and Murad E (1994). Schwertmannite, A new iron oxyhydroxysulphate from Pyhäsalmi, Finland, and other localities. *Mineralogical Magazine* 58: 641–648.

Bigham JM, Schwertmann U, Carlson L and Murad E (1990). A poorly crystallized oxyhydroxysulfate of iron formed by bacterial oxidation of Fe(II) in acid mine waters. *Geochimica et Cosmochimica Acta* 54: 2743–2758.

Bigham JM, Schwertmann U, Traina SJ, Winland RL and Wolf M (1996). Schwertmannite and the chemical modeling of iron in acid sulphate waters. *Geochimica et Cosmochimica Acta* 60: 2111–2121.

Boily J, Gassman PL, Peretyazhko T, Szanyi J and Zachara JM (2010). FTIR spectral components of schwertmannite. *Environmental Science & Technology* 44: 1185–1190.

Brady KS, Bigham JM, Jaynes WF and Logan TJ (1986). Influence of sulphate on Fe-oxide formation: Comparisons with a stream receiving acid mine drainage. *Clays and Clay Minerals* 34: 266–274.

Burton ED, Bush RT, Sullivan LA., Johnston SG and Hocking RK (2008). Mobility of arsenic and selected metals during re-flooding of iron- and organic-rich acid-sulphate soil. *Chemical Geology* 253: 64–73.

Burton ED and Johnston SG (2012). Impact of silica on the reductive transformation of schwertmannite and the mobilization of arsenic. *Geochimica et Cosmochimica Acta* 96: 134–153.

Cao Q, Chen C, Li K, Sun T, Shen Z and Jia J (2021). Arsenic(V) removal behaviour of schwertmannite synthesized by $KMnO_4$ rapid oxidation with high adsorption capacity and Fe utilization. *Chemosphere* 264: 128398.

Carlson C, Bigham JM, Schwertmann U, Kyek A and Wagner F (2002). Scavenging of As from acid mine drainage by schwertmannite and ferrihydrite: A comparison with synthetic analogues. *Environmental Science &Technology* 36: 1712–1719.

Choppala G and Burton ED (2018). Chromium(III) substitution inhibits the Fe(II)-accelerated transformation of schwertmannite. *Plos One* 13: e0208355.

Cruz-Hernández P, Carrero S, Pérez-López R, Fernandez-Martinez A, Lindsay MBJ, Dejoie C and Nieto JM (2019). Influence of As(V) on precipitation and transformation of schwertmannite in acid mine drainage-impacted waters. *European Journal of Mineralogy* 31: 237–245.

Cutting RS, Coker VS, Fellowes JW, Lloyd JR and Vaughan DJ (2009). Mineralogical and morphological constraints on the reduction of Fe(III) minerals by *Geobacter sulfurreducens*. *Geochimica et Cosmochimica Acta* 73: 4004–4022.

Davidson LE, Shaw S and Benning LG (2008). The kinetics and mechanism of schwertmannite transformation to goethite and hematite under alkaline conditions. *American Mineralogist* 93: 1326–1337.

Dey A, Singh R and Purkait MK (2014). Cobalt ferrite nanoparticles aggregated schwertmannite: A novel adsorbent for the efficient removal of arsenic. *Journal of Water Process Engineering* 3: 1–9.

Dong S, Dou X, Mohan D, Pittman Jr. CU and Luo J (2015). Synthesis of graphene oxide/schwertmannite nanocomposites and their application in Sb(V) adsorption from water. *Chemical Engineering Journal* 270: 205–214.

Eskandarpour A, Onyango MS, Ochieng A and Asai S (2008). Removal of fluoride ions from aqueous solution at low pH using schwertmannite. *Journal of Hazardous Materials* 152: 571–579.

Fan C, Guo C, Chen M, Huang W, Wan J, Reinfelder JR and Li X, Zeng Y, Lu G and Dang Z (2019a). Transformation of cadmium-associated schwertmannite and subsequent element repartitioning behaviors. *Environmental Science and Pollution Research* 26: 617–627.

Fan C, Guo C, Zeng Y, Tu Z, Ji Y, Reinfelder JR, Chen M, Huang W, Lu G, Yi X and Dang Z (2019b). The behaviour of chromium and arsenic associated with redox transformation of schwertmannite in AMD environment. *Chemosphere* 222: 945e953.

Fernandez-Martinez A, Timon V, Roman-Ross G, Cuello GJ, Daniels JE and Ayora C (2010). The structure of schwertmannite, a nanocrystalline iron oxyhydroxysulfate. *American Mineralogist* 95: 1312–1322.

Fortin D and Langley S (2005). Formation and occurrence of biogenic iron-rich minerals. *Earth Science Reviews* 72: 1–19.

Gan M, Sun S, Zheng Z, Tang H, Sheng J, Zhu J and Liu X (2015). Adsorption of Cr(VI) and Cu(II) by $AlPO_4$ modified biosynthetic schwertmannite. *Applied Surface Science* 356: 986–997.

Gramp JP, Wang H, Bigham JM, Jones FS and Tuovinen OH (2009). Biogenic synthesis and reduction of Fe(III)-hydroxysulfates. *Geomicrobiology Journal* 26: 275–280.

Hedrich S,Lünsdorf H, Kleeberg R, Heide G, Seifert J and Schl€omann M (2011). Schwertmannite formation adjacent to bacterial cells in a mine water treatment plant and in pure cultures of *Ferrovum myxofaciens*. *Environmental Science & Technology* 45: 7685–7692.

Hockridge JG, Jones F, Loan M and Richmond WR (2009). An electron microscopy study of the crystal growth of schwertmannite needles through oriented aggregation of goethite nanocrystals. *Journal of Crystal Growth* 311: 3876–3882.

Houngaloune S, Hiroyoshi N and Ito M (2015). Effect of Fe(II) and Cu(II) on the transformation of schwertmannite to goethite under acidic condition. *International Journal of Chemical Engineering and Applications* 6: 32–37.

Houngaloune S, Hiroyoshi N and Ito M (2017). Effect of reaction temperature on schwertmannite synthesis, from simulated copper heap leach solutions. *International Journal of Management Information Technology and Engineering* 5: 27–34.

Houngaloune S, Kawaai T, Hiroyoshi N and Ito M (2014). Study on schwertmannite production from copper heap leach solutions and its efficiency in arsenic removal from acidic sulphate solutions. *Hydrometallurgy* 147–148: 30–40.

Huang S and Zhou X (2012). Fe^{2+} oxidation rate drastically affect the formation and phase of secondary iron hydroxysulfate mineral occurred in acid mine drainage. *Material Science and Engineering* C 32: 916–921.

Janneck E, Arnold I, Koch T, Meyer J, Burghardt D and Ehinger S (2010). Microbial synthesis of schwertmannite from lignite mine water and its utilization for removal of arsenic from mine waters and for production of iron pigments. In: Wolkersdorfer C and Freund A (Eds),

Mine water and innovative thinking. IMWA, Sydney, pp. 83–86.

Johnston SG, Burton ED and Moon EM (2016). Arsenic mobilization is enhanced by thermal transformation of schwertmannite. *Environmental Science &Technology* 50: 8010–8019.

Kawano M and Tomita K (2001). Geochemical modeling of bacterially induced mineralization of schwertmannite and jarosite in sulfuric acid spring water. *American Mineralogist* 86: 1156–1165.

Khamphila K, Kodama R, Sato T and Otake T (2017). Adsorption and post adsorption behaviour of schwertmannite with various oxyanions. *Journal of Minerals and Materials Characterization and Engineering* 5: 90–106.

Kirby CS and Brady JAE (1998). Field determination of Fe^{2+} oxidation rates in acid mine drainage using a continuously-stirred tank reactor. *Applied Geochemistry* 13: 509–520.

Kumpulainen S, Kammer F and Hofmann T (2008). Humic acid adsorption and surface charge effects on schwertmannite and goethite in acid sulphate waters. *Water Research* 42: 2051–2060.

Lazaroff N, Melanson L, Lewis E, Santoro N and Pueschel C (1985). Scanning electron microscopy and infrared spectroscopy of iron sediments formed by *Thiobacillus ferrooxiduns*. *Geomicrobiolgy Journal* 4: 231–268.

Lazaroff N, Sigal W and Wasserman A (1982). Iron oxidation and precipitation of ferric hydroxysulfates by resting *Thiobacillus ferrooxidans* cells. *Applied Environmental Microbiology* 43: 924–938.

Li J, Xie Y, Lu G, Ye H, Yi X, Reinfelder JR, Lin Z and Dang Z (2018). Effect of Cu(II) on the stability of oxyanion-substituted schwertmannite. *Environmental Science and Pollution Research* 25: 15492–15506.

Liao Y, Zhou L, Liang J, Xiong H (2009). Biosynthesis of schwertmannite by *Acidithiobacillus ferrooxidans* cell suspensions under different pH condition. *Materials Science and Engineering C* 29: 211–215.

Liao YH, Liang JR and Zhou LX (2011). Adsorptive removal of As(III) by biogenic schwertmannite from simulated As-contaminated groundwater. *Chemosphere* 83: 295–301.

Liu F, Zhou J, Zhang S, Liu L, Zhou L and Fan W (2015). Schwertmannite synthesis through ferrous ion chemical oxidation under different H_2O_2 supply rates and its removal efficiency for arsenic from contaminated groundwater. *Plos One* 10: 1–14.

Loan M, Hart RD, Cowley JM and Parkinson GM (2004). Evidence on the structure of synthetic schwertmannite. *American Mineralogist* 89: 1735–1742.

Loan M, Richmond WR and Parkinson GM (2005). On the crystal growth of nanoscale schwertamannite. *Journal of Crystal Growth* 275: e1875–e1881.

Ma Y and Lin C (2012). Arsenate immobilization associated with microbial oxidation of ferrous ion in complex acid sulphate water. *Journal of Hazardous Materials* 217–218: 238–245.

Mačingová E and Luptáková A (2014). Recovery of iron from acid mine drainage in the form of oxides. *Journal of the Polish Mineral Engineering Society* 15: 193–198.

Maillot F, Morin G, Juillot F, Bruneel O, Casiot C, Ona-Nguema G, Wang Y, Lebrun S, Aubry E, Vlaic G and Brown Jr. GE (2013). Structure and reactivity of As(III)- and As(V)-rich schwertmannites and amorphous ferric arsenate sulphate from the Carnoule's acid mine drainage, France: Comparison with biotic and abiotic model compounds and implications for As remediation. *Geochimica et Cosmochimica Acta* 104: 310–329.

Mazzetti L and Thistlethwaite PJ (2002). Raman spectra and thermal transformations of ferrihydrite and schwertmannite. *Journal of Raman Spectroscopy* 33: 104–111.

Morin G, Juillot F, Casiot C, Bruneel O, Personné J, Elbaz-Poulichet F, Leblang M, Ildefonse P and Calas G (2003). Bacterial formation of tooeleite and mixed arsenic(III) or arsenic(V)-iron(III) gels in the Carnoule's acid mine drainage, France. A XANES, XRD, and SEM study. *Environmental Science & Technology* 37: 1705–1712.

Ooi K, Sonoda A, Makita Y and Torimur M (2017). Comparative study on phosphate adsorption by inorganic and organic adsorbents from a diluted solution. *Journal of Environmental Chemical Engineering* 5: 3181–3189.

Paikaray S, Göttlicher J and Peiffer S (2011). Removal of As(III) from acidic waters using schwertmannite: Surface speciation and effect of synthesis pathway. *Chemical Geology* 283: 134–142.

Pállová Z, Kupka D, and Achimovičová M (2010). Metal mobilization from AMD sediments in connection with bacterial iron reduction. *Mineralia Slovaca* 42: 343–347.

Qiao X, Liu L, Shi J, Zhou L, Guo Y, Ge Y, Fan W, Liu F (2017). Heating changes bio-schwertmannite microstructure and arsenic(III) removal efficiency. *Minerals* 7: 1–14.

Rao SR, Finch JA and Kuyucak N (1995). Technical note ferrous-ferric oxidation in acidic mineral process effluents: Comparison of methods. *Minerals Engineering* 8: 905–911.

Regenspurg S and Peiffer S (2002). A FTIR spectroscopical study to explain bonding structures of arsenate and chromate associated with schwertmannite. In: Schulz HD and Hadeler A (Eds), *Geochemical processes in soil and groundwater*. Wiley-VCH, pp. 78–92.

Regenspurg S and Peiffer S (2005). Arsenate and chromate incorporation in schwertmannite. *Applied Geochemistry* 20: 1226–1239.

Regenspurg S, Brand A and Pfeiffer S (2004). Formation and stability of schwertmannite in acidic mining lakes. *Geochimica et Cosmochimica Acta* 68: 1185–1197.

Reichelt L and Bertau M (2015). Production of ferrihydrite and schwertmanniteusing a microjet mixer device. *Chemical Engineering Research and Design* 98: 70–80.

Schoepfer VA, Burton ED and Johnston SC (2019). Contrasting effects of phosphate on the rapid transformation of schwertmannite to Fe(III) (oxy)hydroxides at near-neutral pH. *Geoderma* 340: 115–123.

Schoepfer VA, Burton ED, Johnston SC and Kraal P (2017). Phosphate-imposed constraints on schwertmannite stability under reducing conditions. *Environmental Science &Technology* 51: 9739–9746.

Schwertmann U and Carlson L (2005). The pH-dependent transformation of schwertmannite to goethite at 25°C. *Clay Minerals* 40: 63–66.

Song J, Jia SY, Ren HT, Wu SH and Han X (2015). Application of a high-surface-area schwertmannite in the removal of arsenate and arsenite. *International Journal of Environmental Science and Technology* 12: 1559–1568.

Stumm W and Morgan J (1995). *Aquatic chemistry: Chemical equilibria and rates in natural waters*. 3rd Ed. Wiley. ISBN: 978-0-471-51185-4.

Šubrt J, Boháčk J, Štengl V, Grygar T and Bezdička P (1999). Uniform particles with a large surface area formed by hydrolysis of $Fe_2(SO_4)_3$ with urea. *Materials Research Bulletin* 34: 905–914.

Šubrt J, Michalková E, Boháček J, Lukáč J, Gánovská Z and Máša B (2011). Uniform particles formed by hydrolysis of acid mine drainage with urea. *Hydrometallurgy* 106: 12–18.

Tresintsi S, Simeonidis K, Vourlias G, Stavropoulos G and Mitrakas M (2012). Kilogram-scale synthesis of iron oxy-hydroxides with improved arsenic removal capacity: Study of Fe(II) oxidation-precipitation parameters. *Water Research* 46: 4255–4267.

Vithana CL, Johnston SG and Dawson N (2018). Divergent repartitioning of copper, antimony and phosphorus following thermal transformation of schwertmannite and ferrihydrite. *Chemical Geology* 483: 530–543.

Wan J, Guo C, Tu Z, Zeng Y, Fan C, Lu G and Dang Z (2018). Microbial reduction of Cr(VI)-loaded schwertmannite by *Shewanella oneidensis* MR-1. *Geomicrobiology Journal* 35: 727–734.

Wang H, Bigham JM and Tuovinen OH (2006). Formation of schwertmannite and its transformation to jarosite in the presence of acidophilic iron-oxidizing microorganisms. *Material Science and Engineering* C 26: 588–592.

Wang X, Gu C, Feng X and Zhu M (2015). Sulfate local coordination environment in schwertmannite. *Environmental Science &Technology* 49: 10440–10449.

Wang Y, Gao M, Huang W, Wang T, Liu Y (2020). Effects of extreme pH conditions on the stability of As(V)-bearing schwertmannite. *Chemosphere* 251: 126427.

Xiong H, Lu X and Wang R (2020). Anionic effect on nanostructure and morphology of bio-schwertmannite dynamically produced within cellular reproduction. *Nanomaterials and Nanotechnology* 10: 1–10.

Ying H, Feng X, Zhu M, Lanson B, Liua F and Wang X (2020). Formation and transformation of schwertmannite through direct Fe^{3+} hydrolysis under various geochemical conditions. *Environmental Science Nano* 7: 2385.

Yu J, Park M and Kim J (2002). Solubilities of synthetic schwertmannite and ferrihydrite. *Geochemical Journal* 36: 119–132.

Zhang S, Jia S, Yu B, Liu Y, Wu S and Han X (2016). Sulfidization of As(V)-containing schwertmannite and its impact on arsenic mobilization. *Chemical Geology* 420: 270–279.

Zhou P, Li Y, Shen Y, Lan Y and Zhou L (2012). Facilitating role of biogenetic schwertmannite in the reduction of Cr(VI) by sulphide and its mechanism. *Journal of Hazardous Materials* 237–238: 194–198.

Zhu F, Guo Z and Hu X (2020). Fluoride removal efficiencies and mechanism of schwertmannite from $KMnO_4/MnO_2$–Fe(II) processes. *Journal of Hazardous Materials* 397: 122789.

4

COMPOSITION AND PROPERTIES OF SCHWERTMANNITE

Their characterization

4.1 Introduction

Schwertmannite possesses unique physical and chemical characteristics in terms of its colour, morphology, spectral positions, surface area, and chemical compositions. A variety of tools and instrumental techniques have been used to characterize schwertmannite for both surficial properties and structural behaviours. In this chapter, we are discussing the most common physical and chemical properties of schwertmannite and the corresponding instrumental tools for their determination.

4.2 Colour

The colour of schwertmannite can vary slightly depending on factors such as the synthesis method, the presence of impurities (e.g., As, Cu, Pb), its origin (natural or synthetic), and its degree of alteration (whether it is partly transformed or pure schwertmannite). Scheinost and Schwertmann (1999) described schwertmannite as having a dark orange-yellow colour when identifying iron minerals in various soil samples. The Munsell colour was recorded with a median value of 8.5 YR, ranging from 6.2 YR to 0.3 Y. The different colours and corresponding Munsell colour indices exhibited by schwertmannite under various environmental conditions are summarized in Table 4.1.

Schwertmannite commonly exhibits a range of colours depending on the conditions of its formation, whether in mine drainage environments or through laboratory synthesis. These colours include reddish-brown (Jiang

DOI: 10.1201/9781003583875-4

TABLE 4.1 Different colours of schwertmannite as observed under different environment of occurrence

Nature of schwertmannite	*Formation medium*	*Colour*	*Munsell colour*	*Sources*
Libiola mine, Italy	Natural	Reddish yellow	5 YR 7/8	Dilelli and Tateo (2002)
Chinkuashih mine, N. Taiwan	Natural	Reddish yellow to reddish brown	–	Jiang and Chen (2005)
Pyhäsalmi sulphide mine, Finland	Natural	Brownish yellow	8 YR	Bigham et al. (1994)
Lomnice mine, Czech Republic	Natural	Yellowish brown	9.0–10 YR	Murad and Rojik (2003)
Coal and metal mines, USA and Finland	Natural	Yellowish brown	9.0–10.0 YR	Bigham et al. (1990)
Libiola mine, Italy	Natural	Yellowish brown	10 YR 5/8	Marescotti et al. (2012)
Lake Matsuo-Goshikinuma, Japan	Natural	–	7.5 YR 6/8	Childs et al. (1998)
Lake Matsuo-Goshikinuma, Japan	Natural	–	10 YR 7/8	Childs et al. (1998)
Matsuo mine drainage treatment plant, Japan	Natural	–	10 YR 7/8	Childs et al. (1998)
Donghae mine, S. Korea	Natural	Brownish yellow	–	Kim et al. (2002)
Dalsung mine, S. Korea	Natural	Brownish or reddish	–	Lee and Kim (2008)
SidiKamber Pb–Zn mine effluents, Algeria	Natural	Brownish red	–	Boukhalfa and Chaguer (2012)
A. ferrooxidans LX5 biosynthesis	Laboratory	Reddish brown	–	Qiao et al. (2017)
Bacterial *synthesis*	Laboratory	Brownish yellow	–	Paikaray and Peiffer (2012)
A. ferrooxidans LX5 biosynthesis	Laboratory	Yellowish	2.5 YR 4/8	Bai et al. (2012)
A. ferrooxidans LX5 biosynthesis	Laboratory	Yellowish	2.5 YR 4/8	Xiong et al. (2008)
A. ferrooxidans	Laboratory	Yellowish red	–	Bigham et al. (2013)
Dialysis	Laboratory	Yellowish brown	8.9 YR 6.5/8.8	Carlson et al. (2002)
Urea *hydrolysis* synthesis	Laboratory	Brownish yellow	–	Eskandarpour et al. (2008)

"–" represents data not available.

and Chen, 2005; Davidson et al., 2008; Qiao et al., 2017), reddish-yellow (Dinelli and Tateo, 2002; Jiang and Chen, 2005), orange (Murad and Rojik, 2003; Burgos et al., 2012; Baleeiro et al., 2018), yellowish-brown (Bigham et al., 1990, 1996; Carlson et al., 2002), brownish-yellow (Paikaray and Peiffer, 2012; Eskandarpour et al., 2008), and yellow (Xiong et al., 2008; Bai et al., 2012) (Figure 4.1). As a result, schwertmannite does not have a single definitive colour, as its hue can vary significantly due to slight changes in composition, the presence of impurities, and the degree of transformation.

For instance, pure schwertmannite synthesized by the dialysis method is typically yellowish-brown, but when 1–10 mM of Cu(II) was introduced, the colour shifted to reddish-brown (Antelo et al., 2013). Similarly, precipitates collected from streams impacted by acid mine drainage (AMD) in SE Ohio, USA, were yellowish, but as the sulphate content exceeded 500 mg L^{-1}, the precipitates showed a reddening effect (Brady et al., 1986).

Experimental handling during the same synthesis method, such as stirring, can result in schwertmannite with varying colours. Baleeiro et al. (2018) observed red-orange schwertmannite when reagent mixing was vigorous before dialysis, while yellow-orange schwertmannite formed when no stirring was performed. This colour difference is reflected by the Fe/S ratio of the products, which were 8.24 in the stirred sample and 4.98 in the unstirred one. Also the nature of the neutralization process of the Fe^{2+} containing solution affects the colour. Burgos et al. (2012) examined four different neutralization processes. When NaOH was added in day 0 (pH 5.18) and day 3 (pH 8.34), schwertmannite formed with an orange colour that remained unchanged until four days of experiment. In contrast, when NaOH was added in day 2 (pH 5.36) and day 3 (pH 6.21), the initial precipitate was dark green, transitioning to orange and orange-brown after one hour. The green colour was attributed to Fe^{2+}-rich and $Fe(OH)_2$-rich water, which created a temporary reducing environment until sufficient oxygen was available to oxidize Fe^{2+} to Fe^{3+}, producing the characteristic orange-red colour.

Foreign ions, such as chloride (Cl^-), also significantly affect the colour of schwertmannite. For example, a reddish-brown schwertmannite with the composition $Fe_8O_8(OH)_{4.68}(SO_4)_{1.66}$ released sulphate when the Cl^-/SO_4^{2-} ratio was adjusted to 3 during Fe^{2+} oxidation by *Acidithiobacillus ferrooxidans* at pH 3.0 (Xiong et al., 2020) with the composition $Fe_8O_8(OH)_{5.6}(SO_4)_{1.2}$, yet maintaining its colour. However, when the Cl^-/SO_4^{2-} ratio was increased to 6 and 10, the colour changed to yellow-orange, with corresponding compositions $Fe_8O_8(OH)_{5.8}(SO_4)_{1.1}$ and $Fe_8O_8(OH)_{5.84}(SO_4)_{1.08}$.

FIGURE 4.1 Mine site images showing different schwertmannite colours. (a) Precipitates from derelict Ottery As–Sn mine in eastern Australia (ref. Schoepfer and Burton, 2021); (b) iron mounts of coal mine drainage, Pennsylvania, USA (ref. Burgos et al., 2012).

4.3 Composition

Schwertmannite precipitation follows a specific stoichiometry with respect to iron (Fe) and sulphur (S). While the ideal schwertmannite has an Fe/S molar ratio of 8, in natural environments or laboratory conditions, this ratio varies between 4.6 and 8.0 (Bigham et al., 1990, 1994, 1996). Sulphate in schwertmannite can be present both as a surface-bound entity and as a structural component, typically varying between 10 and 15 wt%, controlling Fe/S ratios within this range. As a result, the generalized formula $Fe_8O_8(OH)_{8-2x}(SO_4)_x{\cdot}nH_2O$ has been proposed, where $1 \leq x \leq 1.75$. Table 4.2 lists Fe and S contents of several schwertmannite samples and their corresponding chemical formulas obtained from both laboratory studies as well as from natural environments.

The Fe/S ratio can vary significantly depending on the geochemical conditions, the presence of trace metals (e.g., Al, Si, Mg, Ca), and the synthesis pathway. For example, Houngaloune et al. (2015) reported an Fe/S ratio of 5.3 ± 0.35 (Fe = 8.6 ± 0.3; S = 1.62 ± 0.06 mmol g^{-1}) when synthesizing schwertmannite from 10 mM H_2SO_4 and 50 mM Fe^{3+} by titrating with 1 M Na_2CO_3. Caraballo et al. (2013) compiled Fe/S ratios ranging from 3.1 to 10.6. This range can be explained by the dynamic interaction between schwertmannite and dissolved sulphate. Sulphate can constitute up to 35% of the surface-bound fraction, making it easily replaceable or desorbable. Higher Fe/S ratios indicate a lower sulphate content, corresponding to inner-sphere complex binding mechanisms, while lower Fe/S ratios suggest excess sulphate, typically representing outer-sphere complexes due to increased surface adsorption (Fernandez-Martinez et al., 2010; Paikaray and Peiffer, 2010; Baleeiro et al., 2018).

The final sulphate content of schwertmannite is influenced by the synthesis method, dissolved sulphate concentration, and the availability of other ions. Rapid synthesis through H_2O_2 oxidation of Fe^{2+} generally yields schwertmannite with higher Fe/S ratios compared to the dialysis method. Even slight variations in the same synthesis method can result in different Fe/S ratios and chemical compositions. For instance, Baleeiro et al. (2018) synthesized two schwertmannites using the dialysis method with identical reagents, but differences in stirring and cooling rates produced schwertmannite with Fe/S ratios of 8.24 (red-orange) and 4.98 (yellow-orange), respectively. The presence of Cl^- also influences the Fe/S ratio, increasing it from 4.46 to 7.42 as the Cl^-/SO_4^{2-} ratio increased from 1 to 10 during microbial oxidation of 0.1 M $FeSO_4$ by *A. ferrooxidans* (Xiong et al., 2008). This increase in Fe/S ratio is attributed to the replacement of SO_4^{2-} by Cl^-, with the decreasing Fe content reflecting the unavailability of sufficient sulphate for stoichiometric Fe–S complexation.

TABLE 4.2 Iron and sulphur concentrations and Fe/S ratio values of various schwertmannite specimens with their corresponding chemical formulae

Nature of schwertmannite	*Fe (mmol g⁻¹ or wt%)*	*S (mmol g⁻¹ or wt%)*	*Fe/S ratio*	*Chemical formula*	*Sources*
Synthesis from 50 mM Fe^{3+} at 65°C	8.6 ± 0.3 mmol g^{-1}	1.62 ± 0.06 mmol g^{-1}	5.3	$Fe_8O_8(OH)_{4.98}(SO_4)_{1.51}$	HoungAloune et al. (2014, 2015)
Synthesis from 50 mM Fe^{3+} at 65°C	9.11 mmol g^{-1}	1.59 mmol g^{-1}	5.7	$Fe_8O_8(OH)_{5.20}(SO_4)_{1.40}$	Houngaloune et al. (2017)
Synthesis from 50 mM Fe^{3+} at 65°C	8.31 mmol g^{-1}	1.57 mmol g^{-1}	5.3	$Fe_8O_8(OH)_{4.98}(SO_4)_{1.51}$	Houngaloune et al. (2017)
Dialysis	8.40 mmol g^{-1}	1.02 mmol g^{-1}	8.24	$Fe_8O_8(OH)_{6.05}(SO_4)_{0.97}$	Baleeiro et al. (2018)
Dialysis	8.88 mmol g^{-1}	1.79 mmol g^{-1}	4.96	$Fe_8O_8(OH)_{4.77}(SO_4)_{1.61}$	Baleeiro et al. (2018)
Synthesis by H_2O_2 oxidation	8.1 mmol g^{-1}	1.63 mmol g^{-1}	4.97	–	Choppal and Burton (2018)
Synthesis by H_2O_2 oxidation	7.4 mmol g^{-1}	1.5 mmol g^{-1}	4.9	$Fe_8O_8(OH)_{4.96}(SO_4)_{1.65}$	Schoepfer et al. (2017)
Synthesis by H_2O_2 oxidation	42.7 wt%	6.0 wt%	4.05	$Fe_8O_8(OH)_{4.06}(SO_4)_{1.97}$	Wang et al. (2020)
Synthesis by H_2O_2 oxidation	7.55 mmol g^{-1}	1.59 mmol g^{-1}	4.75	$Fe_8O_8(OH)_{4.66}(SO_4)_{1.67}$	Vithana et al. (2015)
Synthesis by H_2O_2 oxidation	7.6 mmol g^{-1}	2.2 mmol g^{-1}	3.45	$Fe_8O_8(OH)_{3.36}(SO_4)_{2.32}$	Fan et al. (2019b)
Synthesis by H_2O_2 oxidation	7.9 mmol g^{-1}	1.9 mmol g^{-1}	–	$Fe_8O_8(OH)_{4.2}(SO_4)_{1.9}\cdot 10H_2O$	Burton et al. (2010)
Synthesis by H_2O_2 oxidation	6.0 mmol g^{-1}	2.4 mmol g^{-1}	–	$Fe_8O_8(OH)_{1.6}(SO_4)_{3.2}$	Ke et al. (2024)
Kristineberg Zn–Cu mine, northern Sweden	8.7 mmol g^{-1}	1.6 mmol g^{-1}	5.4	$Fe_8O_8(OH)_{5.02}(SO_4)_{1.49}\cdot 0.5H_2O$	Jönsson et al. (2005)

(*Continued*)

TABLE 4.2 (Continued)

Nature of schwertmannite	*Fe (mmol g⁻¹ or wt%)*	*S (mmol g⁻¹ or wt%)*	*Fe/S ratio*	*Chemical formula*	*Sources*
Monte Romero mine, SW Spain	38.04 wt%	5.01 wt%	–	$Fe_8O_8(OH)_{4.32}(SO_4)_{1.84}\cdot n\ H_2O$	Acero et al. (2006)
Pyhäsalmi sulphide mine, Oulu, Finland	43.8 wt%	5.09 wt%	4.9	$Fe_8O_8(OH)4_{.8}(SO_4)_{1.6}\cdot 5H_2O$	Bigham et al. (1994)
Smolník mine, Slovak Republic	44.89 wt%	5.83 wt%	4.7	$Fe_{16}O_{16}(OH)_{10}(SO_4)_3\cdot 10H_2O$	Páľlová et al. (2010)
Imgok coal mine drainage, S. Korea	47.07 wt%	5.89 wt%	4.61	$Fe_8O_8(OH)_{4.52}(SO_4)_{1.74}\cdot 8.17\ H_2O$	Yu et al. (1999)
Imgok coal mine drainage, S. Korea	40.7 wt%	5.61 wt%	4.30	$Fe_8O_8(OH)_{4.28}(SO_4)_{1.86}\cdot 8.62H_2O$	Yu et al. (1999)
Monte Romero mine, IPB, SW Spain	41.9 wt%	5.5 wt%	7.62	$Fe_8O_8(OH)_4.34(SO_4)_{1.83}$	French at al. (2012)
Touro Cu mine in Galicia, NW Spain	4.52 mmol g^{-1}	1.44 mmol g^{-1}	3.71	–	Baleeiro et al. (2018)
Touro Cu mine in Galicia, NW Spain	6.77 mmol g^{-1}	1.18 mmol g^{-1}	5.74	–	Baleeiro et al. (2018)
Coal mine drainage, Pennsylvania, USA	–	–	5.42 ± 0.83	$Fe_8O_8(OH)_5(SO_4)_{1.5}$	Burgos et al. (2012)
Lake Matsuo-Goshikinuma, Japan	46.5 wt%	4.06 wt%	6.6	–	Childs et al. (1998)
Lake Matsuo-Goshikinuma, Japan	40.7 wt%	4.52 wt%	5.2	–	Childs et al. (1998)

(Continued)

TABLE 4.2 (Continued)

Nature of schwertmannite	*Fe (mmol g⁻¹ or wt%)*	*S (mmol g⁻¹ or wt%)*	*Fe/S ratio*	*Chemical formula*	*Sources*
Nuestra Señora del Carmen mine pit lake, IPB, Spain	8.59 mmol g^{-1}	1.4 mmol g^{-1}	6.13	$Fe_8O_8(OH)_{5.46}(SO_4)_{1.27}$	Santofimia et al. (2015)
Nuestra Señora del Carmen mine pit lake, IPB, Spain	8.31 mmol g^{-1}	1.63 mmol g^{-1}	5.11	$Fe_8O_8(OH)_{4.92}(SO_4)_{1.54}$	Santofimia et al. (2015)
Nuestra Señora del Carmen mine pit lake, IPB, Spain	8.38 mmol g^{-1}	1.72 mmol g^{-1}	4.88	$Fe_8O_8(OH)_{4.76}(SO_4)_{1.62}$	Santofimia et al. (2015)
Smolník mine, Slovak Republic	44.89 wt%	5.53 wt%	4.7	–	Pállová et al. (2010)
Bacterial oxidation by *A. ferrooxidans*	43.2 wt%	5.17 wt%	4.8	$Fe_8O_8(OH)_{4.7}(SO_4)_{1.65}·6H_2O$	Bigham et al. (1990)
Bacterial oxidation of Fe^{2+}	40 wt%	5.01 wt%	6.3	$Fe_8O_8(OH)_{4.64}(SO_4)_{1.68}$	Paikaray and Peiffer (2012)
Bacterial oxidation of Fe^{2+}	7.4 mmol g^{-1}	1.55 mmol g^{-1}	4.7	$Fe_8O_8(OH)_{6.32}(SO_4)_{1.68}$	Paikaray and Peiffer (2010)
Bacterial oxidation by *A. ferrooxidans*	–	–	5.14	$Fe_8O_8(OH)_{4.89}(SO_4)_{1.55}$	Bai et al. (2012)
Bacterial oxidation by *A. ferrooxidans*	48.5 wt%	3.0 wt%	4.46	$Fe_8O_8(OH)_4·42(SO_4)_{1.79}$	Xiong et al. (2008)
$Fe_2(SO_4)_3·xH_2O$ at 85°C heating for 1 hour	46.1 wt%	5.07 wt%	5.19	$Fe_8O_8(OH)_{4.92}(SO_4)_{1.54}$	Kumpulainen et al. (2008)

Various dissolution techniques, such as ammonium oxalate (~0.2 M) and hydrochloric acid (~5 M), are commonly used to determine the Fe and S content of schwertmannite. The incorporation of trace metals, such as arsenic (As), also alters schwertmannite's Fe and S content. Wang et al. (2020) reported an Fe/S molar ratio of 4.05 for schwertmannite synthesized through rapid H_2O_2 oxidation of Fe^{2+}, with the formula $Fe_8O_8(OH)_{4.06}(SO_4)_{1.97}$ (Fe = 42.7 wt%, S = 6.0 wt%). This ratio increased to 4.96 and 5.27 with the incorporation of 3.8 and 1.9 wt% As(V) through co-precipitation and adsorption, resulting in formulas $Fe_8O_8(OH)_{4.06}(SO_4)_{1.61}(AsO_4)_{0.52}$ and $Fe_8O_8(OH)_{4.06}(SO_4)_{1.52}(AsO_4)_{0.26}$, respectively.

4.4 Density

The true density of schwertmannite is difficult to determine due to its small crystal size. However, based on unit cell dimensions and formulas, the calculated density ranges from 3.77 to 3.99 g cm^{-3} (Bigham et al., 1994). The unit cell parameters used for this calculation were a = 10.66(4) Å, c = 6.04(1) Å, and V = 686(6) $Å^3$, with Z = 1. These calculated densities align with the values reported by Bigham et al. (1990) of 3.75–3.90 g cm^{-3}, using Fe/S molar ratios of 8.0 and 5.3. A later study by Dakos et al. (2012) proposed a lower density of 2.86 ± 0.02 g cm^{-3}.

4.5 Specific surface area

The surface of schwertmannite is characterized by fine needle-like structures that typically radiate from condensed, rounded cores. These surface needles are largely responsible for the high specific surface area (SSA) of schwertmannite. In general, naturally precipitated schwertmannite exhibits an SSA ranging from 125 to 225 m^2 g^{-1}, while synthetic schwertmannite tends to have a higher SSA, between 240 and 320 m^2 g^{-1} (Bigham et al., 1990). However, schwertmannite from mine drainage sites has been reported with lower SSAs, such as ~80 m^2 g^{-1} (Acero et al., 2006), 55 m^2 g^{-1} (Webster et al., 1998), 42.9 m^2 g^{-1} (Jönsson et al., 2005), and 4.54 m^2 g^{-1} (Dakos et al., 2012).

Due to these variations, generalizing SSA for schwertmannite is challenging, as it is heavily influenced by factors such as the formation pathway (bacterial vs. abiotic), the degree of alteration, and the extent of purity. Table 4.3 provides a summary of reported SSAs for different schwertmannite samples.

The SSA of schwertmannite varies significantly depending on synthesis parameters such as pH, temperature, equilibration time, synthesis method (e.g., dialysis, microbial oxidation, H_2O_2 oxidation), the nature of the synthesizing medium (e.g., Milli-Q water vs. As-rich medium), and ionic ratios. Key factors influencing SSA are discussed below.

TABLE 4.3 Specific surface area ($m^2\ g^{-1}$) of schwertmannites that were formed in natural environment or were synthesized in the laboratory by biotic or abiotic synthesis method

Locality/method of synthesis	*SSA ($m^2\ g^{-1}$)*	*Sources*
Smolník abandoned mine, East Slovakia	4.54	Dakos et al. (2012)
Kristineberg Zn–Cu mine, northern Sweden	42.9	Jönsson et al. (2005)
Sunny Corner Ag–Pb–Zn, NSW, Australia	125–200	Chen et al. (2022)
Rapid Fe^{2+} oxidation by H_2O_2	4.0–14.0	Regenspurg et al. (2004)
Rapid Fe^{2+} oxidation by H_2O_2	5.3	Paikaray et al. (2011)
Rapid Fe^{2+} oxidation by H_2O_2	3.2	Li et al. (2011)
Rapid Fe^{2+} oxidation by H_2O_2	48.18	Song et al. (2015)
Rapid Fe^{2+} oxidation by H_2O_2	7.5	Fan et al. (2019a, 2019b)
Rapid Fe^{2+} oxidation by H_2O_2	1.54	Zhang et al. (2016)
Rapid Fe^{2+} oxidation by H_2O_2	24	As et al. (2024)
Dialysis	325.5	Song et al. (2015)
Dialysis	130.9	Dong et al. (2015)
Dialysis	171	Antelo et al. (2012)
Dialysis	161.5	Cutting et al. (2012)
Dialysis	184	Carlson et al. (2002)
Dialysis	177.5	Peiffer and Gade (2007)
Thermal hydroxylation of $Fe_2(SO_4)_3{\cdot}xH_2O$ at 85°C for 1 hour	130	Kumpulainen et al. (2008)
Slow neutralization of 0.01/0.02 mol kg^{-1} SO_4^{2-}	55	Webster et al. (1998)
Microjet mixing	5.6	Reichelt and Bertau (2015)
Continuous flow reactor	168	Tresintsi et al. (2012)
Bacterial oxidation	14.7	Paikaray and Peiffer (2015)
Bacterial oxidation of $FeSO_0{\cdot}7H_2O$ at pH 3.0 with *T. ferrooxidans*	288	Bigham et al. (1996)
Bacterial oxidation of 0.1 M $FeSO_4$ and 0.1 M SO_4^{2-} by *A. ferrooxidans* at pH ~ 2.5	5.27	Xiong et al. (2008)
Bacterial oxidation by *A. ferrooxidans* LX5 at pH 2.5 and 28°C	325.52	Wu et al. (2012)
Bacterial Fe(II) oxidation by *A. ferrooxidans* at pH 3.0	33.52	Bai et al. (2012)
Bacterial oxidation	132.88	Yan et al. (2017)
Urea hydrolysis of mine drainage effluents at 60°C for 72 hours	21.6	*Šubrt et al. (2011)*
Urea hydrolysis	162	Eskandarpour et al. (2008)
$Fe_2(SO_4)_3$ addition to 85°C preheated Milli-Q H_2O	42.3	Carrero et al. (2017)
$FeCl_3$ and Na_2SO_4 method (without dialysis)	206.1	Dou et al. (2013)
$FeCl_3$ and Na_2SO_4 method (without dialysis)	120	Xiong et al. (2023)

Abbreviation: SSA, specific surface area.

4.5.1 *Synthesis method*

SSA of schwertmannite differs based on the synthesis method used. Schwertmannite produced by rapid H_2O_2 oxidation of Fe^{2+} typically has the lowest SSA, ranging from ~4.0–5.5 $m^2\ g^{-1}$ (Burton et al., 2008, 2010; Paikaray et al., 2011) to 48.18 $m^2\ g^{-1}$ (Song et al., 2015). In contrast, schwertmannite synthesized by the dialysis method has the highest SSA, from ~200–300 $m^2\ g^{-1}$ (Bigham et al., 1990; Paikaray et al., 2011) to 325.52 $m^2\ g^{-1}$ (Song et al., 2015). Bacterial oxidation also results in high SSA, e.g., 288 $m^2\ g^{-1}$ for schwertmannite formed through the oxidation of Fe^{2+} at pH 3 by *Thiobacillus ferrooxidans* (Bigham et al., 1990).

4.5.2 *Synthesis parameters*

4.5.2.1 pH

Solution pH has been shown to positively affect SSA. For example, schwertmannite formed by bacterial oxidation at pH 1.6 had an SSA of 3.42 $m^2\ g^{-1}$, while at pH 3.44, the SSA increased to 23.45 $m^2\ g^{-1}$ (Liao et al., 2011). Similar findings were reported by Tresintsi et al. (2012), where SSA increased from 53 $m^2\ g^{-1}$ at pH 3 to 155 $m^2\ g^{-1}$ at pH 5.5 in a continuous flow reactor. When the precipitate formed at pH 5.5 was dialyzed for 30 days, the SSA increased further to 168 $m^2\ g^{-1}$.

4.5.2.2 Temperature

Higher synthesis temperatures tend to increase SSA. For instance, Kumpulainen et al. (2008) reported an SSA of 130 $m^2\ g^{-1}$ for schwertmannite synthesized at 85°C for 1 hour. Houngaloune et al. (2017) observed a tenfold increase in SSA by raising the synthesis temperature from 25°C to 65°C, with SSA values ranging from 14.1–21.4 $m^2\ g^{-1}$ to 147.4–176.9 $m^2\ g^{-1}$. Cutting et al. (2009) found that schwertmannite synthesized by ramping the temperature of 10 mM $Fe_2(SO_4)_3{\cdot}9H_2O$ to 70°C achieved an SSA of 203.6 $m^2\ g^{-1}$ compared to 159.4 $m^2\ g^{-1}$ for the same synthesis via the dialysis method.

4.5.2.3 Equilibration time

Longer equilibration times can increase SSA. For example, Reichelt and Bertau (2015) reported an increase in SSA from 5.6 ± 1.2 $m^2\ g^{-1}$ to 35 ± 1 $m^2\ g^{-1}$ after equilibrating schwertmannite in its parent suspension for one week, likely due to the development of micropores and mesopores.

4.5.2.4 Rate of Fe^{2+} oxidation

The rate of Fe^{2+} oxidation affects SSA, with the rapid oxidation by H_2O_2 generally resulting in smaller particle size and lower SSA. For instance, Liu et al. (2015) found that adding 1.8 mL of H_2O_2 over 2 hours produced particles with a size of 680 nm and an SSA of 2.06 $m^2\ g^{-1}$, while slower H_2O_2 addition (0.36 mL over 50 hours) resulted in larger particles (1.81 μm) and a higher SSA of 16.3 $m^2\ g^{-1}$. Zhang et al. (2019) reported similar findings, where slower H_2O_2 addition rates (0.5 mL min^{-1}) increased SSA from 169.31^2 to 228.75 $m^2\ g^{-1}$.

4.5.3 Synthesis medium

4.5.3.1 Nature of Fe^{2+} source

The type of Fe^{2+} salt used in synthesis influences SSA. Schwertmannite synthesized from $FeCl_2$ had an SSA of 196 $m^2\ g^{-1}$, compared to 168 $m^2\ g^{-1}$ for $FeSO_4$ at pH 5.5 (Tresintsi et al., 2012). The lower SSA for $FeSO_4$ may be due to sulphate adsorption on schwertmannite surfaces, inhibiting complete N_2 adsorption.

4.5.3.2 Availability of contaminants and foreign ions

The presence of contaminants affects SSA, depending on the sorption potential of schwertmannite for specific elements. For example, Fan et al. (2019a) observed little change in SSA for schwertmannite formed in the presence of 0.225 g L^{-1} $CdSO_4$ (7.7 $m^2\ g^{-1}$ vs. 7.5 $m^2\ g^{-1}$ for pure schwertmannite). In contrast, Paikaray and Peiffer (2012, 2015) reported an increase in SSA from 14.7 to 17.1 $m^2\ g^{-1}$ when 0.92 wt% As(III) was incorporated. Similarly, Cutting et al. (2012) found that As(V) enrichment led to SSA increases from 161.5 to 212.2 $m^2\ g^{-1}$ as As(V) content rose from 0.3 to 4.2 wt%.

Ionic ratios, such as As/Fe and Cl^-/SO_4^{2-}, also influence SSA. Zhang et al. (2016) found that as the As/Fe ratio increased, SSA increased from 1.53 to 7.2 $m^2\ g^{-1}$. Similarly, Xiong et al. (2008) observed an increase in SSA from 5.27 to 14.2 $m^2\ g^{-1}$ as the Cl^-/SO_4^{2-} ratio rose from 0 to 10 during microbial oxidation.

Bigham et al. (1990) reported that schwertmannite's SSA varies systematically with the sulphate content. SSA increased from 250 $m^2\ g^{-1}$ at low SO_4^{2-} concentrations to 350 $m^2\ g^{-1}$ at 750 mg L^{-1}, but decreased as SO_4^{2-} concentration exceeded 1500 mg L^{-1}. Kumpulainen et al. (2008) found that humic acid also influences SSA, increasing it from 130 to 261 $m^2\ g^{-1}$

as humic acid concentration rose, though excessive humic acid (100 mg C L^{-1}) reduced SSA to 152 m^2 g^{-1} due to surface site blockage.

4.6 Point of zero charge (pH_{PZC})

The point of zero charge (pH_{PZC}) of schwertmannite is the pH at which its surface charge is neutral. This property plays a significant role in the retention of heavy metals and metalloids. Positively charged contaminants (e.g., Cd^{2+}, Cu^{2+}) are retained at pH > pH_{PZC}, where the surface charge of schwertmannite is negative. Inversely, negatively charged contaminants (e.g., $H_2AsO_4^-$, PO_4^{3-}) are typically retained at pH < pH_{PZC}. Thus, pH_{PZC} is a crucial parameter in controlling the fate of contaminants in mine drainage and polluted areas. In some studies, the parameter isoelectric point is used, which is more or less identical to the pH_{pzc}. In order to make these studies comparable with studies in which pH_{pzc} was determined, we consistently used the term pH_{pzc} in our overview.

Natural iron oxides generally have a lower pH_{PZC} due to impurities. Surface-adsorbed anionic species, such as CO_3^{2-} or SO_4^{2-}, reduce surface charge and thus also the pH_{pzc}, while cationic species, such as Cu^{2+} or Zn^{2+}, increase it.

Measured pH_{PZC} values for schwertmannite vary depending on the synthesis method and medium. Schwertmannite synthesized by dialysis generally has a higher pH_{PZC} (e.g., 7.2, Jönsson et al., 2005) than those synthesized by bacterial mediation (e.g., 5.4, Liao et al., 2011). Schwertmannite formed by urea neutralization exhibits the lowest pH_{PZC} values, around 3.6–4.2 (Eskandarpour et al., 2008; Goswami and Purkait, 2014). Adsorption of substances during the biosynthesis method possibly lowers pH_{PZC}. Table 4.4 provides a compilation of pH_{PZC} values from the literature.

Baleeiro et al. (2018) derived pH_{pzc} (IEP in this study) values from titrimetric measurements for schwertmannite formed from mine waters to

TABLE 4.4 Point of zero charge (pH_{PZC}) values for schwertmannites and their respective synthesis method

pH_{PZC}	*Synthesis method*	*Sources*
7.2	Dialysis method	Jönsson et al. (2005)
7.0	Dialysis method	Khamphila et al. (2017)
5.1	Dialysis method	Dong et al. (2015)
5.4	Bacterial synthesis	Liao et al. (2011)
3.1	Bacterial synthesis	Wu et al. (2012)
4.2	Urea hydrolysis	Eskandarpour et al. (2008)
3.6	Urea hydrolysis	Goswami and Purkait (2015)

be approximately 10.2, compared to 7.5–8.5 determined for laboratory-synthesized schwertmannite using the dialysis method. The mine water from the Touro copper mine in Galicia, NW Spain, contained high concentrations of Cu, Al, Zn, and Ni, unlike the dialysis method. Apparently, the presence of high concentrations of positively charged cations adsorbing to the surface led to a shift to the high pH_{pzc} value of 10.2. Similar observations were made at high loading with the anion SO_4^{2-}. Specimens with low pH_{pzc} had a higher SO_4^{2-} content (1.79 mmol g^{-1}) as compared to specimens with high pH_{pzc} (1.02 mmol g^{-1}). Kumpulainen et al. (2008) reported that the pH_{pzc} (IEP in this study) for schwertmannite synthesized by heating an iron solution at 85°C for 1 hour decreased from 6.9–8.2 to 2.5 when the synthesis medium contained 100 mg C L^{-1} of humic acid anions.

4.7 Fractal geometry

Fractal dimensions are used to analyse the surface and structural irregularities of schwertmannite, which are linked to its adsorption behaviour. This analysis typically involves electron microscopy at various scales and N_2 adsorption–desorption behaviour (Zhang et al., 2017). A higher fractal dimension (Df) indicates a more complex pore structure, which can hinder diffusion and desorption of pollutants.

The study used two types of schwertmannite: (1) bacterially synthesized schwertmannite by *A. ferrooxidans* LX5 and (2) schwertmannite synthesized by the dialysis method. The calculated Df was slightly lower for dialytically synthesized schwertmannite (Df = 2.6) compared to bacterially synthesized schwertmannite (Df = 2.7), suggesting that schwertmannite from different origins has different fractal dimensions. Schwertmannite has two fractal dimensions: (1) surface Df and (2) inner Df. The component with the higher Df (surface or inner) dominates the overall adsorption process. A higher surface Df indicates that surface processes control pollutant uptake, particularly for larger pollutants (e.g., humic acid), while a higher inner Df suggests that the structure of the schwertmannite governs the adsorption of smaller pollutants (e.g., Cr(VI).

4.8 Morphology

The morphology of schwertmannite has been extensively studied using scanning electron microscopy (SEM) and high-resolution transmission electron microscopy (HRTEM). In SEM studies, schwertmannite samples are typically mounted on aluminium stubs with double-sided carbon tape and coated with gold or carbon to enhance conductivity (Loan et al., 2004; Gan et al., 2016). The current intensity during SEM analysis varies depending on the instrument used. For TEM analysis, the preparation involves placing a small amount of the washed and dried sample (~0.01 g) in an ultrasonic bath for 30 minutes in 5 mL of ice-cooled Milli-Q water. A few drops of this suspension are placed on a holey carbon TEM grid and air-dried. Energy-dispersive spectra (EDS) are collected at a rate of 2000–7000 counts per second for 50 seconds (Loan et al., 2004).

Prior to the formal recognition of schwertmannite as a mineral, SEM micrographs of precipitates formed through bacterial oxidation of 0.2 M $FeSO_4$ by *T. ferrooxidans* revealed a fibroporous microstructure. This precipitate, referred to as "amorphous ferric hydroxysulfate", was reddish-brown and highly soluble in 4N HCl, exhibiting characteristics consistent with schwertmannite (Lazaroff et al., 1982). Although this phase was associated with jarosite in Na^+, NH_4^+, and K^+-containing media, both crystalline jarosite with surface spines and acicular microstructures were observed. The absence of X-ray diffraction (XRD) analysis in these early studies left the identity of the material unconfirmed, but it was later recognized as schwertmannite.

Schwertmannite is generally characterized by a spheroidal morphology with acicular spines, sometimes referred to as fine needles or whiskers, measuring approximately 60–100 nm in length and 2–10 nm in width (Bigham et al., 1990, 1996; Regenspurg et al., 2004; Loan et al., 2004, 2005; Paikaray et al., 2011; Paikaray and Peiffer, 2012; Marescotti et al., 2012). These spines project outward from an electron-dense core of rounded aggregates, which range from 200 to 500 nm in diameter. Various researchers have used a different terminology to describe the morphology of schwertmannite. It has been referred to as "pin-cushion", "hedge-hog", "filamentous", or "ball-and-whisker" morphology (Bigham and Nordstrom, 2000; Loan et al., 2004; Kim and Kim, 2021). These distinctive features are used as qualitative identifiers to differentiate schwertmannite from other amorphous precipitates (Table 4.5). Hence, a sharp distinction among these morphological features is not established.

For example, Hockridge et al. (2009) described "hedge-hog" morphology, with rounded aggregates a few hundred nanometres in diameter and radiating whiskers of 2–10 nm wide. Davidson et al. (2008) observed "pin-cushion" morphology, with aggregates of 500–1000 nm in diameter and spines of 5–10 nm wide and 10–50 nm long. Johnston et al. (2016) reported a similar diameter range of ~400–900 nm spheroidal morphology. Cutting et al. (2009) described schwertmannite as having ribbon-like filaments forming outwardly radiating spherical "ball-and-whisker" structures, with whiskers of 50–70 nm long and 3–6 nm wide, exhibiting a polycrystalline nature due to the irregular arrangement of lattice fringes within individual filaments.

Most schwertmannite specimens described so far exhibit small, needle-shaped crystals densely aggregated into larger particles, typically ranging from 200 to 500 nm in diameter, characterized by unusual "pin-cushion" morphology (Bigham et al., 1996; Burton et al., 2006). This unique morphology, with its surface spines often referred to as "hedge-hog" or "pin-cushion" patterns, is considered a defining feature of schwertmannite (Bigham et al., 1994; Bigham and Nordstrom, 2000; Acero et al., 2005, 2006; Jönsson et al., 2006). The distinct "hedge-hog" appearance was also reported by Brady et al. (1986) in an early study, where schwertmannite was initially identified as amorphous ferric oxides. This was confirmed through TEM, which revealed "finger-like projections" from the surface. The same morphology has been observed in natural schwertmannite samples collected from AMD-affected stream beds in SE Ohio, USA. These natural samples closely resemble synthetic schwertmannite precipitated from a 0.02 M $Fe(NO_3)_3$ solution with 1500 mg L^{-1} SO_4^{2-} (as Na_2SO_4) (Figure 4.2).

The spheroidal particles of schwertmannite consist of needles, though it is challenging to determine their exact origin. Careful observation reveals that these particles often begin with more than two needles at the centre, which merge into a single needle as they grow outward (French et al.,

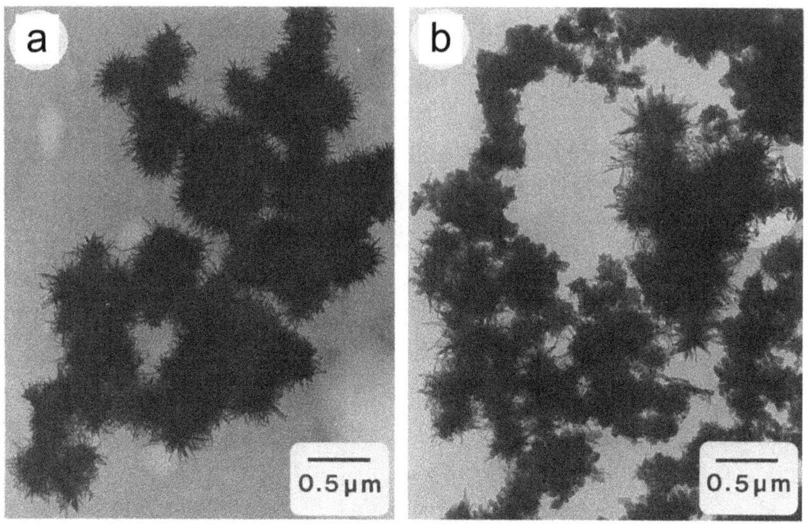

FIGURE 4.2 Transmission electron micrographs of (a) water-dialyzed specimens synthesized at $c(SO_4^{2-})$ = 1500 mg L^{-1} and (b) the natural stream precipitate from acid mine drainage, SE Ohio, USA (ref. Brady et al., 1986). Note the similarity in morphology of the two specimens despite of differences in formation conditions.

2012). The commonly observed "hedge-hog" morphology is traditionally thought to develop around bacterial cells due to nucleation and crystallite growth. However, studies indicate that it can form under abiotic conditions as well (Götz, 2011).

It is important to note that the characteristic "pin-cushion" or "hedge-hog" morphology is not always present, especially in field samples, as reported by Jönsson et al. (2005) and Antelo et al. (2013). Peretyazhko et al. (2009) similarly observed the absence of typical filamentous surface features in schwertmannites formed from mine drainage effluents, which instead developed large aggregates up to ~2 μm in diameter. These aged products maintained their spherical morphology, even when the mineralogy transformed into goethite, consistent with the findings of Jönsson et al. (2005).

Morphology alone can be insufficient for accurately characterizing schwertmannite, as aggregates of ferrihydrite with surface spines can closely resemble schwertmannite, especially when considering that both minerals are amorphous and stable under acidic conditions (Hockridge et al., 2009). Research has shown that ferric oxyhydroxide precipitates can resemble the "hedge-hog" morphology of schwertmannite when viewed under low magnification. However, XRD and HRTEM at higher magnification reveal that these aggregates are actually ferrihydrite (Loan et al., 2005).

4.8.1 Effect of synthesis conditions on schwertmannite morphology

Schwertmannite nanoparticles have also been synthesized through rapid Fe^{2+} oxidation pathways. For example, Zhang et al. (2019) observed "ball-and-whiskers" morphology during the synthesis of schwertmannite via the addition of 30% H_2O_2 to a 60 mM Fe^{2+} solution. The morphology varied depending on the rate of H_2O_2 addition, with slower additions (1 and 0.5 mL min^{-1}) resulting in better-developed "ball-and-whiskers" patterns and increased particle size. Slower oxidation rates were found to improve particle development, leading to the formation of pores and cavities, which increased pore volume and decreased pore size (Figure 4.3).

The synthesis method also affects schwertmannite morphology. For example, rapid oxidation of Fe^{2+} using H_2O_2 produces spherical aggregates without surface spines or needles, while the "hedge-hog" morphology with ~100-nm-long surface spines is observed in schwertmannite synthesized through the dialysis method (Song et al., 2015; Zhang et al., 2018). The diameter of the spherical aggregates also differs, being larger (~500 nm) in the H_2O_2 synthesis compared to the "hedge-hog" pattern (~100 nm).

The diameter of spheroids grows over time during synthesis. For instance, Liao et al. (2011) observed that schwertmannite synthesized

FIGURE 4.3 SEM (a and b) and TEM (d) images showing "pin-cushion" morphology (ref. Bigham et al., 1994; Acero et al., 2006; Xiong et al., 2008, respectively); (c) and (e) SEM/TEM micrographs showing "ball-and-whisker" morphology (ref. Zhang et al., 2019; Cutting et al., 2009, respectively); (f) "hedge-hog" morphology (ref. Loan et al., 2005) of schwertmannite.

by *A. ferrooxidans* LX5 grew from ~500 nm after 1 hour to ~600 nm after 5 hours, reaching 2.5 µm after 72 hours, and finally ~4 µm after 168 hours, remaining stable until 216 hours. Initially, the schwertmannite formed as aggregates of small spheroids without spines or needles, but the "hedge-hog" morphology developed after 72 hours, with needle size increasing over time.

Hockridge et al. (2009) demonstrated the development of highly amorphous precipitates after heating $Fe_2(SO_4)_3 \cdot xH_2O$ for just 5 minutes, with surface whiskers forming after 10 minutes and growing to over 200 nm in length after 120 minutes.

4.8.2 Effect of foreign ions on schwertmannite morphology

Foreign ions like As, Al, and Cr can also affect schwertmannite morphology. For instance, Gan et al. (2015a) found that increasing Al content in the synthesis medium caused the destruction of surface spines, resulting in smoother spheroids and a reduction in particle size to ~500 nm. Similarly, Xiong et al. (2008) reported that high Cl^- concentrations in the medium led to a morphology resembling akaganèite, despite the mineralogy being identified as schwertmannite (Figure 4.4).

The "pin-cushion" morphology of schwertmannite deteriorates when arsenic (As) is incorporated into its structure (Carlson et al., 2002; Paikaray et al., 2011; Park et al., 2016). This incorporation shortens the spine length and roughens the surface morphology, with co-precipitated arsenic causing more pronounced morphological degradation than adsorbed arsenic (Park et al., 2016). Several studies have noted the blurring of schwertmannite's "pin-cushion" morphology based on the As/(As + S) molar ratio. For example, Carlson et al. (2002) observed this effect when the As/(As + S) ratio exceeded 0.4, and the rounded morphology became thinner with porous centres as the ratio increased to 0.6. Park et al. (2016) found similar results at a lower ratio of As/(As + S) = 0.019 (Figure 4.5).

Arsenate incorporation inhibits the nucleation of schwertmannite, preventing the well-developed spines and leading to rough surfaces, particularly at higher As/Fe ratios (Maillot et al., 2013; Zhang et al., 2016). Schwertmannites synthesized by the H_2O_2 method display smooth surfaces, comparable to those formed with arsenic incorporation up to an As/Fe ratio of 0.03, beyond which the surface becomes increasingly irregular (Zhang et al., 2016). In the dialysis method, typical spines are absent at higher As concentrations.

Paikaray and Peiffer (2015) observed slight degradation of the morphology, with shorter spines, after incorporating 0.92 wt% As(III). In contrast, Cutting et al. (2012) found that As incorporation during synthesis

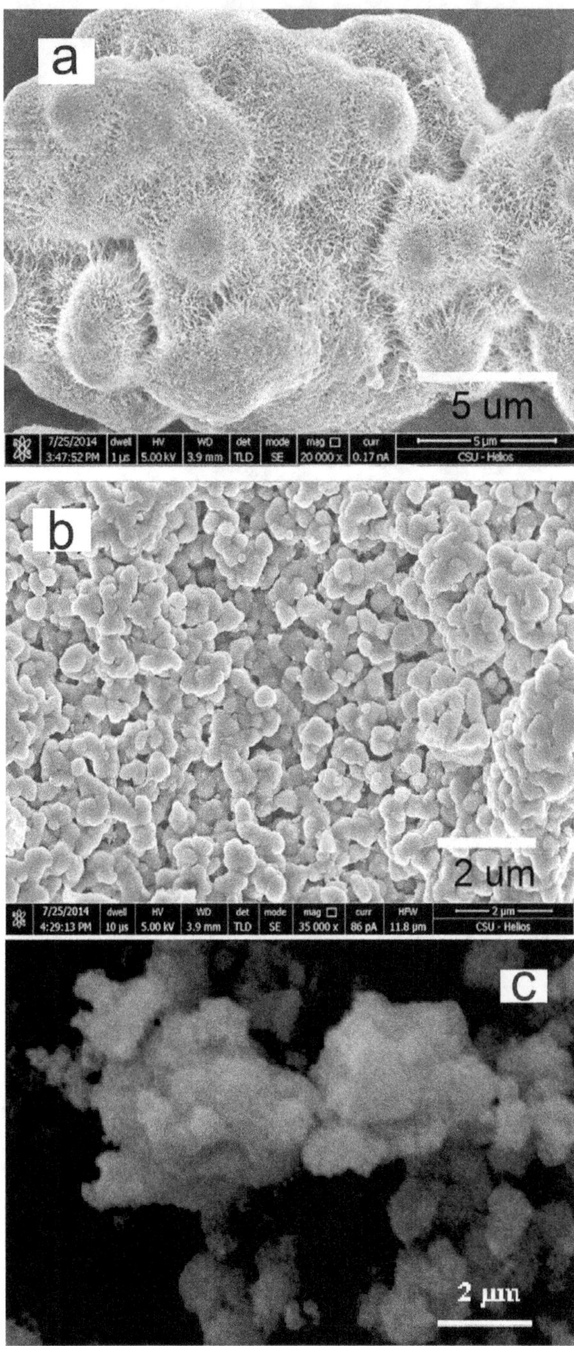

FIGURE 4.4 SEM images of the biosynthesized schwertmannite at $FeSO_4 \cdot 7H_2O$/$AlPO_4$ ratios of 20/0.1 (a) and 20/0.5 (b) (Gan et al., 2015a), and a Cl^-/SO_4^{2-} molar ratio of 10 (c) (Xiong et al., 2008).

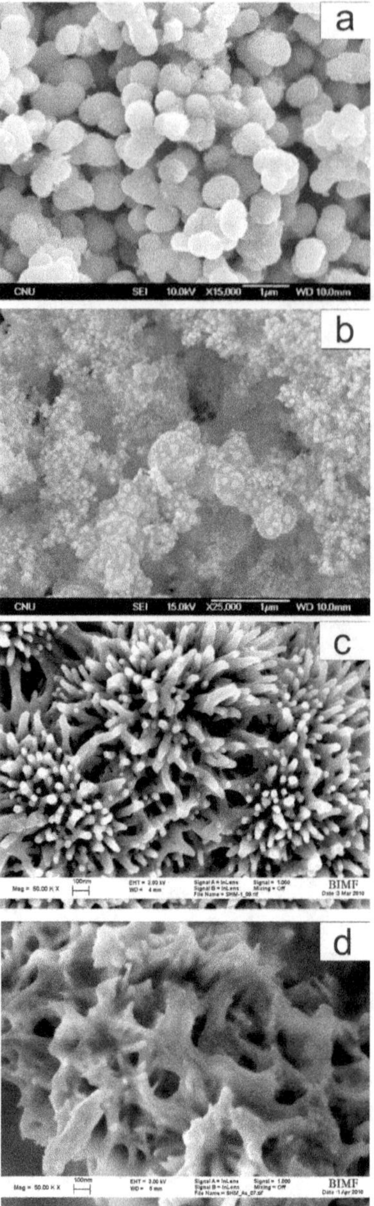

FIGURE 4.5 SEM images of schwertmannites that show the impact of arsenic on morphology. (a) Synthesis by rapid Fe^{2+} oxidation with H_2O_2 in the presence of As(V) being adsorbed, (b) co-precipitated with As(V) (ref. Park et al., 2016), (c) SEM images of schwertmannite synthesized by the dialysis method in the absence and (d) in the presence of 0.92 wt% As-adsorbed schwertmannite (ref. Paikaray and Peiffer, 2015). Note how the morphology gets deteriorated upon arsenic incorporation. More deterioration was visible in case of schwertmannite with co-precipitated As compared to adsorbed As (b vs. a) and schwertmannite containing no As compared to 0.92 wt% As (c vs. d).

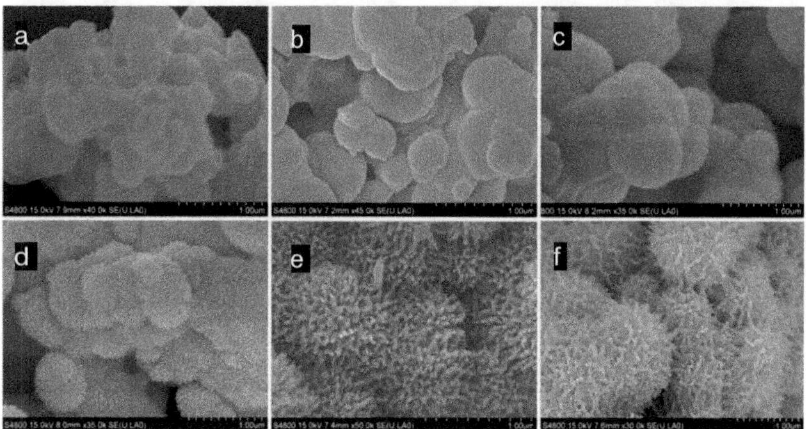

FIGURE 4.6 Field emission SEM images for schwertmannite produced through bacterial oxidation of Fe^{2+} and collected after 1 (a), 4 (b), 8 (c), 16 (d), 24 (e), and 36 (f) hours of incubation (ref. Xiong et al., 2020). Note the development of surface morphology with time from a smooth surface to pin-cushion morphology.

did not cause significant morphological changes, except for a reduction in particle size that inversely correlated with arsenic content. For instance, particle diameter decreased in the order of 4.13 < 2.3 < 0.79 < 0.3 < 0 wt% As. Wang et al. (2020) similarly reported a size reduction from ~500 to ~100 nm with 3.8 wt% As(V) incorporation.

Chromium (Cr) also affects schwertmannite morphology. Choppala and Burton (2018) found that Cr(III) incorporation at 0.02, 0.05, and 0.21 mmol g^{-1} led to spheroidal morphology degradation and smaller particle sizes. Wan et al. (2018) reported a rough schwertmannite morphology when synthesized by the dialysis method, with "hedge-hog" features appearing in the presence of K_2CrO_4 at concentrations of 1 and 10 g L^{-1}, resulting in solid-phase loadings of 0.37 and 0.60 wt%, respectively.

Xiong et al. (2020) studied the temporal evolution of the schwertmannite morphology using field emission SEM. Schwertmannites synthesized via the *A. ferrooxidans* Fe^{2+} oxidation pathway initially showed smooth surfaces until 4 hours of incubation. After 8 hours, the surface became granulated, and the "pin-cushion" morphology developed after 16 hours (Figure 4.6). In contrast, when Cl^- replaced SO_4^{2-} during synthesis, homogeneous spheres with granulated surfaces formed by 4 hours, with "hedge-hog" features emerging after 36 hours along with an increase in particle diameter. Early stage morphology often resembled ferrihydrite, but XRD analysis confirmed the presence of schwertmannite even after 1 hour of incubation, ruling out ferrihydrite as a precursor (Loan et al., 2005; Hockridge et al., 2009).

TABLE 4.5 Types of morphology, particle size, and origin of schwertmannites

Origin of schwertmannite	*Medium of synthesis/locality*	*Type of morphology*	*Particle diameter (nm/µm)*	*Sources*
Synthesized	Hydrothermal method	Hedge-hog	~100 nm	Wang et al. (2013)
Synthesized	Thermal hydroxylation of $Fe_2(SO_4)_3{\cdot}xH_2O$ at 60°C for 2 hours	Pin-cushion	300 nm to 1 µm	Cruz-Hernández et al. (2019)
Synthesized	Urea hydrolysis	Pin-cushion	Needles: 2–4 nm wide; 60–90 nm length	Eskandarpour et al. (2008)
Synthesized	Microbial oxidation by *A. ferrooxidans* LX5	Spherical particles	~2.5 µm	Wu et al. (2012)
Synthesized	Microbial oxidation by *A. ferrooxidans*	Pin-cushion	0.5 µm	Xiong et al. (2023)
Synthesized	Microbial oxidation by *A. ferrooxidans* LX5	Sea urchin and flower	–	Zhang et al. (2017)
Synthesized	Microbial oxidation by *A. ferrooxidans* LX5	Hedge-hog	–	Yan et al. (2017)
Synthesized	Microbial oxidation by *A. ferrooxidans* LX5	Pin-cushion	1–2 µm	Xiong et al. (2008)
Synthesized	Microbial oxidation by *A. ferrooxidans* HX3	Pin-cushion	1.2 µm	Xu et al. (2014)
Synthesized	Dialysis	Pin-cushion	2.5 µm	Xiong et al. (2023)
Synthesized	Dialysis	Sea urchin and flower	–	Ying et al. (2020)
Synthesized	Dialysis	Pin-cushion	200–500 nm	Carlson et al. (2002)
Synthesized	H_2O_2 rapid Fe^{2+} oxidation	Spheroidal particles	1000 nm	Duan et al. (2016)
Synthesized	H_2O_2 rapid Fe^{2+} oxidation	Irregular smooth spheres	1 µm	Fan et al. (2019a)
Synthesized	H_2O_2 rapid Fe^{2+} oxidation	Spheroids	400–600 nm	Choppala and Burton (2018)

(*Continued*)

TABLE 4.2 (Continued)

Synthesized	H_2O_2 rapid Fe^{2+} oxidation	Hedge-hog	–	Mačngová and Luptáková (2014)
Synthesized	H_2O_2 rapid Fe^{2+} oxidation	Ball-with-whiskers	–	Zhang et al. (2019)
Synthesized	H_2O_2 rapid Fe^{2+} oxidation	Pin-cushion	–	As et al. (2024)
Natural	Acidic lowlands of eastern Australia	Pin-cushion	–	Burton et al. (2008)
Natural	Acidic lowlands of eastern Australia	Hedge-hog	500 nm	Burton et al. (2007)
Natural	Chinkuashih mine drainage, N. Taiwan	Hedge-hog	–	Jiang and Chen (2005)
Natural	AMD from Thompson and Leaf Rapids, Manitoba, Canada	Sea urchin and web-like aggregates	–	Sidenko and Sherriff (2005)
Natural	Kristineberg Zn–Cu mine, northern Sweden	Spherical particles	400 nm	Jönsson et al. (2005)
Natural	Smolník abandoned mine, East Slovakia	Pin-cushion	–	Dakos et al. (2012)
Natural	Ilkwang mine, Pusan, Korea	Pin-cushion	–	Park et al. (2016)
Natural	Monte Romero abandoned mine, SW Spain	Pin-cushion	–	Acero et al. (2005, 2006)
Natural	Acid sulphate soils, eastern Australia	Pin-cushion	–	Burton et al. (2006)
Natural	Chinkuashih mine drainage, N. Taiwan	Hedge-hog	–	Jiang and Chen (2005)
Natural	Libiola Fe–Cu sulphide mine, Italy	Pin-cushion with whiskers	–	Marescotti et al. (2012)
Natural	Nuestra Señora del Carmen mine pit lake, IPB, Spain	Hedge-hog	–	Santofimia et al. (2015)
Natural	South Shetlands, Antarctica	Sea urchin	–	Dold et al. (2013)
Natural	Lake 77, Lusatian mining area, Germany	Hedge-hog	–	Miot et al. (2016)
Natural	Dalsung mine, SE Korea	Hedge-hog/pin-cushion	1.5–2.5 μm	Kim and Kim (2021)
Natural/synthetic	Lignite pit Lake 77, Lusatian, E. Germany/*Acidithrix ferrooxidans* strain C25	Pin-cushion	–	Mori et al. (2016)

"–" represents data not available.

4.9 Crystallinity and XRD patterns

Mineralogical identification of schwertmannite is primarily conducted through XRD techniques, which have been widely used to analyse both field and laboratory samples. Depending on the specific requirements of the X-ray diffractometer, different sample preparation methods may be applied. Generally, schwertmannite samples are dried at 40°C–60°C to remove moisture content, then finely powdered using an agate mortar (Asta et al., 2007; Paikaray et al., 2011, 2017). The powdered samples are mounted on sample holders using a suitable mounting liquid, such as ethanol or methanol, and placed in the instrument for X-ray analysis.

Three types of detectors are commonly employed to capture the diffraction peaks of schwertmannite: Co Kα (λ = 1.79 Å), Mo Kα (λ = 0.71 Å), and Cu Kα (λ = 1.54 Å). The operating parameters, such as tube voltage, tube current, and scan rate, vary between different studies. For instance, typical settings include 40 kV and 40 mA, with a scan rate ranging from ½ to 0.15 °2θ min^{-1} (Sidenko and Sherriff, 2005; HoungAloune et al., 2014). The samples are irradiated across a 2θ range of 0–80°, and the diffractograms are recorded using software for further interpretation (Dinelli and Tateo, 2002; Asta et al., 2007; Gan et al., 2016).

4.9.1 *Pure schwertmannite*

Schwertmannite and akaganèite (β-FeOOH) share similar crystal symmetries. Akaganèite forms via the hydrolysis of ferric chloride in the absence of dissolved SO_4^{2-}. However, the presence of SO_4^{2-} together with Cl^- gradually alters the diffraction patterns, with increasing SO_4^{2-} leading to broader peaks as Cl^- is replaced by SO_4^{2-}. The resulting product at high SO_4^{2-} concentrations is the characteristic 8-line diffractogram of schwertmannite (Figure 4.7). Schwertmannite exhibits distinct diffraction peaks, with sharp and prominent reflections at 0.254 and 0.151 nm, which are present in all samples. Other peaks, though weaker and broader, occur at 0.164, 0.172, 0.331, and 0.495 nm. A shoulder peak of 0.254 nm can also be seen around 0.223–0.232 nm. Typically, the eight diffraction peaks are observed at approximately 0.510, 0.345, 0.255, 0.228, 0.195, 0.166, 0.151, and 0.146 nm, corresponding to specific lattice planes (200, 111), (310), (212), (302), (412), (522), (004), and (204, 542), respectively. The most dominant reflections are from the (212) and (004) planes (Davidson et al., 2008).

Bigham et al. (1994) calculated the unit cell parameters for schwertmannite as a = b = 10.66(4) Å, c = 6.04(1) Å, V = 686(6) $Å^3$, with Z = 1. The probable space group is proposed as P4/m. Despite its poor crystallinity

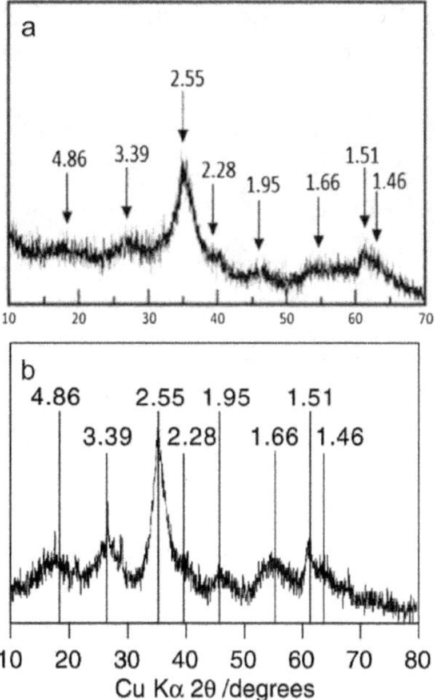

FIGURE 4.7 X-ray diffractograms of (a) schwertmannite synthesized from hydrolysis of 50 mM $Fe_2(SO_4)_2$ at 65°C (ref. HoungAloune et al., 2014) and (b) natural schwertmannite collected from the Kristineberg Zn–Cu mine, northern Sweden (ref. Jönsson et al., 2005). Note the similarity of diffractogram patterns between natural-occurring and laboratory-synthesized schwertmannites.

and partially amorphous structure, schwertmannite consistently displays these characteristic diffraction peaks.

Bigham et al. (1990) were the first to identify eight distinct diffraction peaks for synthetic and naturally precipitated schwertmannite. Both biotic and abiotic schwertmannite samples have been found to align with these eight diffraction lines. Table 4.6 provides a summary of the typical diffraction peak positions reported in various studies, and Figure 4.7 shows some representative diffractograms for schwertmannite.

Several XRD peaks of ferrihydrite overlap with those of schwertmannite in its pure form. The most intense peak for both schwertmannite and ferrihydrite occurs at ~0.254 nm, making it difficult to differentiate between them. However, schwertmannite can be distinguished by the presence of a peak at ~0.164 nm, while ferrihydrite's corresponding peak is slightly displaced at ~0.172 nm. Other overlapping d-spacings include peaks at ~0.228 and ~0.195 nm, which are identical for both minerals and, therefore, hard to distinguish independently. The reflections recorded at 0.151 and 0.145 nm for schwertmannite also overlap with ferrihydrite, but the intensity ratios are reversed, i.e., 0.151 nm peak is intense in schwertmannite compared to 0.145 nm and reverse for ferrihydrite (Dakos et al., 2012). Additionally, two unique reflections at ~0.339 and 0.486 nm are observed in schwertmannite, which are absent in ferrihydrite.

Table 4.6 X-ray diffractogram peaks of schwertmannites from field and laboratory studies

Diffraction peaks (nm)	*hkl*	*Synthesis medium/natural specimens*	*Sources*
0.566, 0.393, 0.296, 0.227, 0.193, 0.176, 0.170	(210), (310), (212), (302), (113), (522), (204)	H_2O_2 rapid Fe^{2+} oxidation	Xie et al. (2017)
0.510, 0.345, 0.255, 0.228, 0.195, 0.166, 0.151, 0.146	–	H_2O_2 rapid Fe^{2+} oxidation	Duan et al. (2016)
0.637, 0.613, 0.553, 0.465, 0.395, 0.352, 0.263, 0.182	–	H_2O_2 rapid Fe^{2+} oxidation	Wang et al. (2020)
0.487, 0.339, 0.256, 0.195, 0.166, 0.151	(210), (310), (212), (113), (522), (004)	H_2O_2 rapid Fe^{2+} oxidation	Zhang et al. (2016)
0.393, 0.296, 0.265, 0.228, 0.193, 0.176	–	H_2O_2 rapid Fe^{2+} oxidation	Fan et al. (2019a)
0.512, 0.336, 0.255, 0.230, 0.194, 0.166, 0.155	–	30 days dialysis	Carlson et al. (2002)
0.486, 0.339, 0.255, 0.228, 0.195, 0.166, 0.151, 0.146	–	H_2SO_4 + Fe^{3+} titration by Na_2CO_3	HoungAloune et al. (2014)
0.495, 0.336, 0.256, 0.225, 0.195, 0.167, 0.152, 0.146	–	85°C heated Fe^{3+} solution	Loan et al. (2004)
0.565, 0.394, 0.296, 0.265, 0.227, 0.193, 0.175, 0.170	–	Bacterial Fe^{2+} oxidation by *A. ferrooxidans* LX5	Yan et al. (2017)
0.456, 0.328, 0.255, 0.219, 0.191, 0.186, 0.151, 0.146	–	Bacterial Fe^{2+} oxidation by *A. ferrooxidans*	Gramp et al. (2008)
0.476, 0.33, 0.256, 0.233, 0.197, 0.167, 0.151, 0.146	–	Bacterial Fe^{2+} oxidation by *A. ferrooxidans*	Wang et al. (2006)

(*Continued*)

TABLE 4.6 (Continued)

Diffraction peaks (nm)	*hkl*	*Synthesis medium/natural specimens*	*Sources*
0.545, 0.330, 0.258, 0.230, 0.193, 0.168, 0.152, 0.146	–	Bacterial oxidation of mine water	Paikaray and Peiffer (2010)
0.509, 0.337, 0.256, 0.228, 0.195, 0.168, 0.151, 0.148	–	Bacterial Fe^{2+} oxidation by *A. ferrooxidans*	Xiong et al. (2020)
0.507, 0.326, 0.254, 0.224, 0.194, 0.167, 0.151	–	Bacterial oxidation of mine water	Paikaray and Peiffer (2012)
0.489, 0.340, 0.256, 0.233, 0.197, 0.168, 0.151, 0.146	–	Bacterial Fe^{2+} oxidation by *T. ferrooxidans* at pH 3	Bigham et al. (1996)
0.520, 0.332, 0.255, 0.277, 0.196, 0.163, 0.151, 0.146	–	Paroistenjärvi mine at Ylőjärvi, Finland	Carlson et al. (2002)
0.499, 0.336, 0.255, 0.227, 0.195, 0.167, 0.151, 0.147	–	Lomnice mine, Czech Republic	Murad and Rojik (2003)
0.486, 0.339, 0.255, 0.228, 0.195, 0.166, 0.151, 0.146	–	Kristineberg Zn–Cu mine, northern Sweden	Jönsson et al. (2005)
0.510, 0.340, 0.255, 0.230, 0.190, 0.170, 0.151, 0.140	–	Luikonlahti mine drainage, Finland	Kumpulainen et al. (2007)
0.479, 0.335, 0.254, 0.228, 0.194, 0.162, 0.151, 0.146	–	Lake Matsuo-Goshikinuma, Japan	Childs et al. (1998)
0.486, 0.339, 0.255, 0.228, 0.195, 0.166, 0.151, 0.146	–	Smolník abandoned mine, Slovak Republic	Pállová, et al. (2010)
0.486, 0.339, 0.255, 0.228, 0.195, 0.166, 0.151, 0.146	(200, 111), (310), (100, 212), (302), (412), (522), (004), (204, 542)	Pyhäsalmi sulphide mine, Oulu, Finland	Bigham et al. (1994)

(*Continued*)

TABLE 4.6 (Continued)

Diffraction peaks (nm)	*hkl*	*Synthesis medium/natural specimens*	*Sources*
0.486, 0.339, 0.255, 0.195, 0.166, 0.151	–	Libiola Fe–Cu sulphide mine, Italy	Marescotti et al. (2012)
0.501, 0.343, 0.253, 0.193, 0.166, 0.150	(210), (310), (212), (412), (522), (004)	Smolník mine, Slovakia	Kupka et al. (2012)
0.486, 0.339, 0.255, 0.228, 0.195, 0.166, 0.151, 0.146	–	Nuestra Señora del Carmen mine pit lake, IPB, Spain	Santofimia et al. (2015)
0.256, 0.225, 0.199, 0.165, 0.150	–	South Shetlands, Antarctica	Dold et al. (2013)

"–" represents data not available.

These peak positions can vary between samples due to differences in purity, composition, alteration effects, and crystallinity. The degree of crystallinity is inferred from the sharpness of the peaks and the signal-to-noise ratio: sharper peaks and better signal-to-noise ratios indicate a higher relative crystallinity. However, schwertmannite is typically an amorphous or poorly crystalline mineral phase. The broad peaks indicate poor crystallinity, but the mineral is not X-ray amorphous. Even with uniform synthesis methods, crystallinity can vary based on parameters such as the rate of H_2O_2 addition. Faster H_2O_2 addition results in noisier patterns and lower diffraction intensities, while slower addition produces clearer peaks with less noise (Zhang et al., 2019).

4.9.2 Effect of foreign ions on XRD patterns

Aluminium substitution during schwertmannite precipitation also affects the diffractogram patterns. As the Fe/Al ratio decreases from 200 to 4, most characteristic schwertmannite peaks diminish, with the peak at 0.256 nm shifting to lower angles (0.310 nm) at high Fe/Al ratios (Gan et al., 2015b). Aluminium, when present in the synthesizing medium, co-precipitates with schwertmannite, especially in Al-enriched mine waters. This co-precipitation lowers the intensity of characteristic peaks, including the strongest peak at 0.256 nm (Baleeiro et al., 2018). Shifts in XRD peaks are also observed during the neutralization of AMD effluents, attributed to the entrapment of Al and trace metals like Cu, Zn, and Ni (Baleeiro et al., 2018). Similarly, incorporation of As(III) into schwertmannite, for example, caused a slight shift in the 0.326 nm peak to 0.338 nm, although other band positions and intensities remained unchanged (Paikaray and Peiffer, 2012). Incorporation of Cu(II) (HoungAloune et al., 2014), As(V) (Zhang et al., 2016), and graphene (Dong et al., 2015) reduced the intensities of the diffractogram peaks. An increase in crystallinity, with a corresponding decrease in Fe/S ratio, has been observed at higher hydrolysis temperatures (25°C–60°C) and longer dialysis times (1–15 days). However, temperatures above 60°C may result in the formation of goethite, which reduces the sulphate content and leads to poorer crystallinity (Ying et al., 2020).

Nevertheless, these shifts and reductions in peak intensity are not always visible. Some studies report no significant changes in peak positions or intensities even with varying contaminant levels. For instance, Choppala and Burton (2018) found that Cr(III) incorporation (0.02–0.21 mmol g^{-1}) did not affect schwertmannite's peak positions at 0.486, 0.339, 0.255, 0.228, 0.195, 0.166, and 0.151 nm. Similarly, studies by Song et al. (2015) and Xiong et al. (2020) found no differences in peak positions for schwertmannite synthesized by different methods or with high Cl^-/SO_4^{2-} ratios.

4.9.3 Mixtures with other ferric oxyhydroxides

In complex mixtures of schwertmannite with other crystalline iron oxides such as goethite and jarosite, differential X-ray diffraction (DXRD) has been used for identification. DXRD assumes that crystalline phases do not dissolve in acidified ammonium oxalate (AAO) medium. Isolation of the XRD signal of schwertmannite is obtained by subtracting the XRD signal of the AAO-treated sample from the untreated sample (Dold and Fontboté, 2001; Vithana et al., 2015). However, DXRD can be limited by the presence of non-target components like organic matter, so it is recommended to use other identification techniques in addition to DXRD.

4.10 Thermal properties

Schwertmannite exhibits a unique thermal response when subjected to temperatures above room temperature, demonstrating both endothermic and exothermic behaviours. The thermal behaviour of schwertmannite is characterized by an initial endothermic reaction between 100°C and 300°C, followed by an exothermic reaction between 540°C and 580°C, and a second endothermic reaction between 650°C and 710°C (Figure 4.8) (Bigham et al., 1990, 1994; Jambor et al., 1995). During the first endothermic reaction, schwertmannite undergoes a weight loss of about 15%–25%,

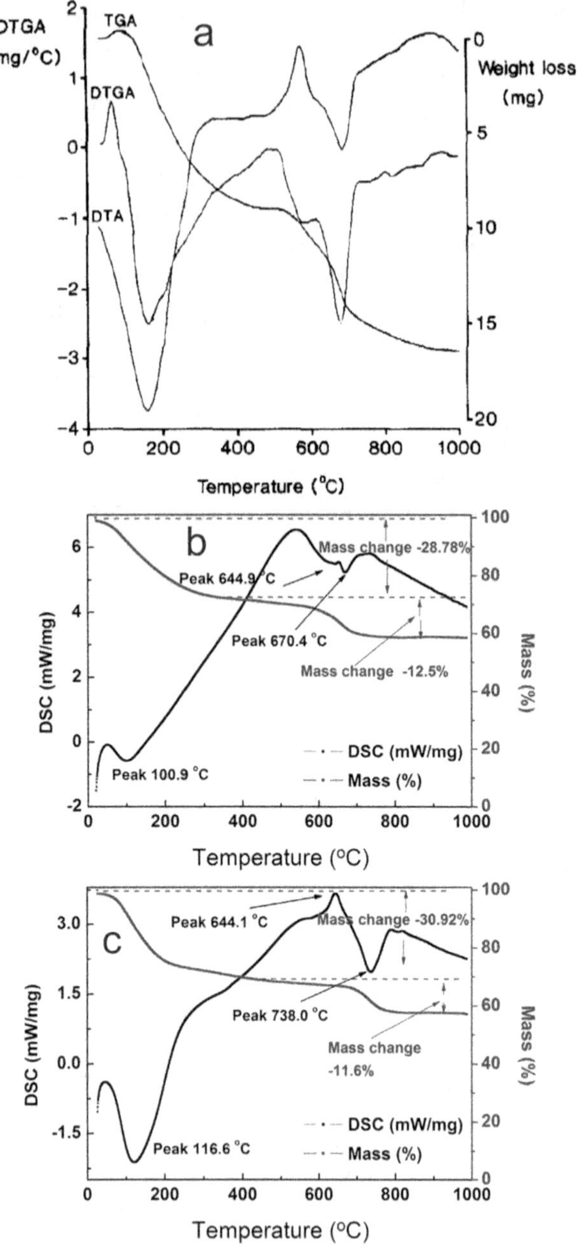

FIGURE 4.8 (a) Differential thermal (DTA), thermogravimetric (TGA), and differential thermogravimetric (DTGA) patterns of schwertmannite (ref. Bigham et al., 1990). TGA pattern of schwertmannite and corresponding mass loss of schwertmannite formed through biooxidation of $FeSO_4 \cdot 7H_2O$ in the presence of $AlPO_4$ at an Fe/Al ratio of (b) 20/0.1 and (c) 20/5 (ref. Gan et al., 2015a).

which can be attributed to the vaporization of sorbed H_2O and structural OH/H_2O. This partial destruction of schwertmannite's structure results in asymmetric XRD patterns, though the peak at ~0.254 nm remains.

As the temperature increases, schwertmannite begins to destabilize and transforms into poorly crystalline hematite around 460°C. This hematite increases in crystallinity as the temperature continues to rise, forming along with solid $Fe_2(SO_4)_3$ just before the exothermic reaction occurs at ~540°C–580°C (Jambor et al., 1995; Mačingová and Luptáková, 2014). The exothermic reaction is believed to result from the crystallization of $Fe_2(SO_4)_3$. As the temperature further increases, $Fe_2(SO_4)_3$ decomposes, contributing to the final endothermic reaction at ~700°C, during which hematite and SO_3 are produced. The release of SO_3 leads to additional weight loss.

The specific temperatures of these endothermic and exothermic reactions can vary depending on the schwertmannite's origin, whether from AMD environments or laboratory synthesis. For instance, Yu et al. (1999) reported two endothermic peaks for schwertmannite from Imgok Creeks, Korea: one at <200°C (due to loss of adsorbed H_2O) and another at ~630°C (attributed to loss of structural SO_4). In Donghae mine, South Korea, Kim et al. (2002) found that the schwertmannite's dehydration was completed at ~500°C with a weight loss of ~23.6%, and the final desulfuration began at ~500°C.

The Fe/Al ratio also influences the thermal behaviour of schwertmannite. Gan et al. (2015b) observed that increasing the Fe/Al ratio from 0.1 to 5 shifted the endothermic peaks from 100.9°C and 670.4°C to 116.6°C and 738°C, indicating improved thermal stability due to Al substitution. Schwertmannites containing ~4.5 wt% As_2O_5 from the Wieoeciszowice and Radzimowice mine sites in Poland also exhibited differences in thermal behaviour. While no significant changes in thermogravimetric analysis (TGA) were observed between pure and As(V)-sorbed schwertmannites, the differential thermal analysis (DTA) peaks for endothermic and exothermic reactions occurred at lower temperatures. The first endothermic peak, related to dehydration and dihydroxylation, appeared between 160°C and 170°C, causing a ~25% weight loss, followed by an exothermic peak at 530°C–550°C due to recrystallization of the anhydrous phase (Parafiniuk and Siuda, 2006). The final endothermic peak, occurring at 655°C–680°C, was associated with the loss of SO_3 and arsenic oxide, resulting in a mass loss of ~11 wt%.

4.11 Vibrational spectroscopical properties—infrared and Raman spectral properties

Fourier transformed infrared (FTIR) spectroscopy is widely used to identify functional groups in schwertmannite by determining their specific spectral band positions. The finely powdered samples are mixed with standards

TABLE 4.7 Fourier transformed infrared absorption bands of schwertmannites

Vibration mode	*Band positions*	*Paikaray and Peiffer (2010)*	*Brady et al. (1986)*	*Boily et al. (2010)*	*Mačingová and Luptáková (2014)*	*Kumpulainen et al. (2007)*	*Parafiniuk and Siuda (2006)*	*Xiong et al. (2008)*	*Bai et al. (2012)*	*Zhang et al. (2019)*	*As et al. (2024)*
ν(OH)	3322–3424	–	–	–	3350	–	3322–3424	3344	3318	–	–
H_2O	1631–1633	1630, 1730	–	–	1640	1622	1631–1633	1700–1400	–	–	–
$\nu_3(SO_4)$	1053–1196	1050, 1120	1035, 1120	1120	1051	1130, 1070, 1125	1053–1196	1124	–	1166, 1128, 1051	1170, 1120, 1050
$\nu_1(SO_4)$	979–988	–	980	981	985	979	979–988	979	980	978	980
δ(OH)	877–880	881	875	–	–	880, 887, 890	877–880, 796–806	–	–	855	860
$\nu_3(AsO_4^{3-})$	820–828	–	–	–	–	828	820–828	–	1127	–	712
Fe–O stretch	667–700	700	690	–	695	700	667–700	682	700	703	–
$\nu_4(SO_4)$	604–612	610	595	610	604	611, 609	604–612	614	612	611	610
Fe–O stretch	408–460	420	–	–	470	424, 422	408–460	–	–	427	425

(e.g., 0.3–2.5 mg of sample + 300 mg KBr as described by Dinelli and Tateo (2002), and pellets are prepared for exposure to infrared (IR) radiation. Typically, 32 scans are performed to minimize noise at a scan speed of about 4 cm^{-1}, although some studies employ higher resolutions (e.g., 1 cm^{-1}) and increased scans (up to 500) depending on the sample (Bigham et al., 1994).

FTIR scans are generally carried out across a range of wavenumbers between 100 and 4000 cm^{-1} to capture the spectral positions corresponding to functional groups involving Fe, S, O, OH, and trace metals, particularly for contaminated schwertmannite samples. The spectral band positions and their corresponding intensities can vary depending on the sample's composition, method of preparation, and instrumentation. Despite these variances, certain spectral positions are consistently observed for pure schwertmannite. Attenuated total reflection–FTIR (ATR-FTIR) is used to understand sulphur speciation in schwertmannite, Fe–O, S–O, Fe–S coordination, and complexation behaviour (inner vs. outer sphere) (Boily et al., 2010; Wang et al., 2015). These key positions, along with their band assignments, are summarized in Table 4.7, and a representative spectrum is provided in Figure 4.9.

Wavenumbers between 2900 and 3700 cm^{-1} are indicative for O–H stretching (ν_{OH}) absorption bands; the large band at 1633 cm^{-1} is assigned to H–O–H deformation (Bigham et al., 1994; Boukhalfa and Chaguer, 2012; Gan et al., 2015a). The broad OH-stretching band at 3300 cm^{-1}

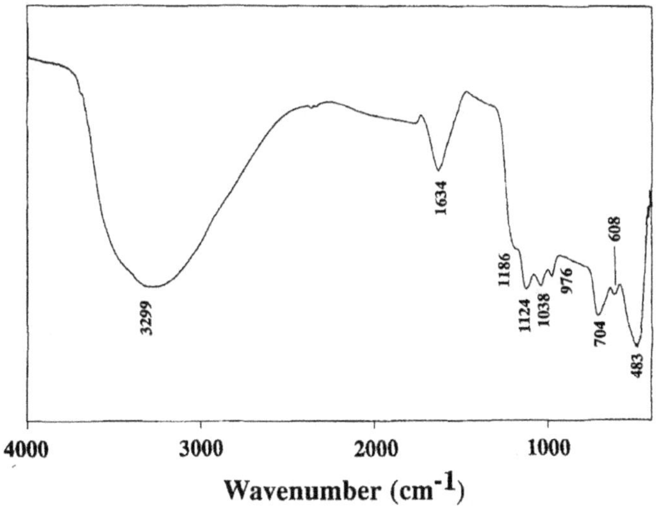

FIGURE 4.9 Infrared spectra of schwertmannite showing its band positions (ref. Bigham et al., 1994).

is considered to be the most intense band (Bigham et al., 1994). This OH-stretching band is also assigned to bands at 3385 and 3200 cm^{-1} (Boily et al., 2010). There are three main forms of hydroxyl groups on the surface of iron oxides, the single coordinated hydroxyl group ($OH^{-1/2}$), the double coordinated hydroxyl group (OH^0), and the triple coordinated hydroxyl group ($OH^{+1/2}$), which combine with one, two, and three iron atoms, respectively. The $OH^{-1/2}$ is the most reactive with arsenate, while OH^0 is chemically inert under normal conditions (Cao et al., 2021).

Sulphate in schwertmannite has different IR absorption features. Three IR bands at 1186, 1124, and 1038 cm^{-1} are assigned to splitting of $v_3(SO_4)$ due to formation of a bidentate bridging complex between SO_4 and Fe. Splitting of $v_3(SO_4)$ vibrations in schwertmannite indicates lowering of symmetry with two bands corresponding to a monodentate complex (C_3v symmetry) and one to a bidentate complex (C_2v symmetry). Such complexes are explained by the occurrence of outer-sphere complexes ($FeOH^{2+}SO_4^{2-}$ or $FeOH^{2+}HSO_4^-$), which approximate the tetrahedral symmetry of free SO_4^{2-} and which form by replacement of OH groups with SO_4 either at the mineral surface through ligand exchange or within the structure during nucleation and subsequent growth of the crystal (Bigham et al., 1990). These complexes are regarded to originate from the physisorbed sulphate ions at the diffuse layer that are stabilized by weak electrostatic interactions or hydrogen bonding (Tresintsi et al., 2014). The lowering of symmetry is also indicated by $v_1(SO_4)$ and $v_4(SO_4)$ vibrations at 979–988 cm^{-1} and 604–612 cm^{-1}, respectively (Bigham et al., 1990; Parafiniuk and Siuda, 2006). The $v_4(SO_4)$ band is characteristic of structural SO_4^{2-} while the appearance of a weak $v_1(SO_4)$ peak is indicative of partial distortion due to outer-sphere complexes (Tresintsi et al., 2014; Zhang et al., 2019). The vibrations at 690–704 and 410–483 cm^{-1} are assigned to Fe–O stretching vibrations. A broad absorption shoulder in the 800–880 cm^{-1} range is apparent in some specimens, which is related to OH deformation (δ(OH)) vibrations.

4.12 Effect of drying conditions on FTIR absorption bands in schwertmannite

Wang et al. (2015) used ATR-FTIR method to understand sulphur coordination of schwertmannite synthesized by dialysis method. A broad triply degenerate asymmetric stretching (v_3) band at ~1105 cm^{-1} with two shoulder bands at ~1040 and 1170 cm^{-1}, a v_1 fundamental of the symmetric sulphate stretching at 980 cm^{-1}, and a v_4 bending band at ~608 cm^{-1} were found in ATR-FTIR spectra for both dried and wet samples that are representative of sulphate inner-sphere complexes. The v_3 band splits to a lower degree for both samples, which suggests less inner-sphere complexes. The decreased intensities of these bands are ascribed to the increasing sulphate loss at higher pHs. Furthermore, the v_3 band splits more for the dried samples than that for the wet samples at a given pH, suggesting that dried samples contain more inner-sphere complexes. The OH (δ_{OH}) deformation (~840 cm^{-1}) and Fe–O stretch bands (685 cm^{-1}) for dried sample and their intensities decrease as pH increased suggesting structural disorder of schwertmannite. However, the δ_{OH} and Fe–O stretch bands shift to lower wavenumbers (~786 and ~665 cm^{-1}, respectively) in wet samples, indicating extensive hydrogen bonding.

4.13 Effect of foreign ions on FTIR absorption bands in schwertmannite

Arsenate-rich schwertmannites exhibit a unique absorption band at 820–828 cm^{-1} that corresponds to v_3 deformations of the AsO_4^{3-} group. Paikaray and Peiffer (2012) found appearance of a new IR spectral band at ~1385 cm^{-1} for schwertmannite that is loaded with 0.92 wt% As(III) while other band positions remained unchanged with no major variation in spectral intensities. Kumpulainen et al. (2007) reported a similar band at ~1399 cm^{-1} with relatively intense reflections for schwertmannite-rich precipitates from several mine drainage ditches, ponds, and waste drainages, which they assigned to COO^- groups. Alternatively, the occurrence of this band may be explained with As(III) enrichment of the precipitates with an As content of up to 1.1 mmol g^{-1}.

The IR absorption band positions of Al^{3+}-containing schwertmannites (up to 0.69 wt%) were identical to that of pure schwertmannites. However, the presence of Al^{3+} increased the intensity of the hydroxyl and sulphate functional groups by broadening the peak widths and depths (Zhu et al., 2020). The intense absorption bands at ~2500–3600 cm^{-1} represent O–H stretching (v_{OH}) and H–O–H deformation band at ~1600 cm^{-1}, whereas a splitting of fundamental band at ~1136 cm^{-1} and absorption band at ~707 cm^{-1} was caused by $v_3(SO_4^{2-})$ and $v_4(SO_4^{2-})$ vibrations, respectively. The bands observed near 607 and 450 cm^{-1} are representative of vibrations of FeO_6 coordination octahedral.

Cao et al. (2021) studied As(V) retention onto schwertmannites those that are synthesized by H_2O_2 and $KMnO_4$ oxidation of $FeSO_4$ under RT and P conditions at pH 3. Triply degenerate asymmetric stretching band (v_3) at 1100 cm^{-1}, a fundamental of the symmetric SO_4^{2-} stretching band (v_1) at 980 cm^{-1}, and a bending band (v_4) at 610 cm^{-1} are common bands aligning with other studies. The 840, 795, and 685 cm^{-1} are ascribed to the OH deformation (δ_{OH}), goethite characteristic (γ_{OH}), and Fe–O stretching bands, respectively. FTIR spectroscopy is also helpful to dismantle competitive sorption of anions. For schwertmannites to which As(V) is adsorbed, band positions were consistent at pH 2, 7, and 11 (Cao et al., 2021). However, the intensities of SO_4^{2-} absorption bands decreased at pH 11 while those of OH vibration bands increased demonstrating the replacement of schwertmannite SO_4^{2-} by OH^- under basic conditions. This observation was further confirmed by the appearance of an OH^- absorption band at 1610 cm^{-1} at pH 11 only and consequent lowering of the As(V) uptake capacity.

Sulphate content in schwertmannite determines the intensities of SO_4^{2-} vibration bands by lowering the 1130, 1040, 970, and 610 cm^{-1} when SO_4^{2-} content of schwertmannite is low. An increase in As content in schwertmannite led to a substantial change of the absorption bands. With increasing As/(As + S) ratio, a characteristic symmetric stretching vibration of an As–O–Fe bond at 840 cm^{-1} gradually appeared, accompanied by a shift of the OH stretching vibration at 3370 cm^{-1} to approximately 3010 cm^{-1} and the diminishing of all sulphate bands (Carlson et al., 2002). When the As/(As + S) ratio exceeded 0.8, also the 700 cm^{-1} absorption band attributed to OH stretching in schwertmannite is lost leaving a residual absorption feature at 657 cm^{-1}.

Raman spectroscopy is used to characterize schwertmannite. It shows characteristic bands at 294, 318, 350, 421, 544, 580, 715, and 981 cm^{-1} (Mazzetti and Thistlethwaite, 2002; Burton et al., 2009; Parviainen et al., 2015). Raman spectral signature of As(V) sorption onto schwertmannite has also been detected at ~850 cm^{-1}, which is assigned to As(V)–O stretching band (Burton et al., 2009; Parviainen et al., 2015).

4.14 Mössbauer parameters

Mössbauer spectroscopy is a widely used technique to characterize schwertmannite and its metastable counterparts, providing insights into the bonding of iron within the schwertmannite structure and its transformation products. Schwertmannite's Fe and O are octahedrally coordinated (Cashion and Murad, 2012), where oxygen atoms can exist as O^{2-} or as part of H_2O, OH^-, or SO_4^{2-} groups. As a result, several Fe–O configurations are possible, each displaying distinct isomer shifts (ISs) and quadrupole splittings (QSs).

Mössbauer spectroscopy has been extensively used to study the magnetic properties and binding behaviour of iron, oxygen, and sulphur in schwertmannite. As temperature decreases, magnetic order becomes more pronounced due to a distribution of Néel temperatures, rather than superparamagnetic relaxation. At room temperature (298 K), schwertmannite exhibits an asymmetric doublet, similar to ferrihydrite, but with greater asymmetry due to the presence of SO_4^{2-} ions (Figure 4.10, Bigham et al., 1990).

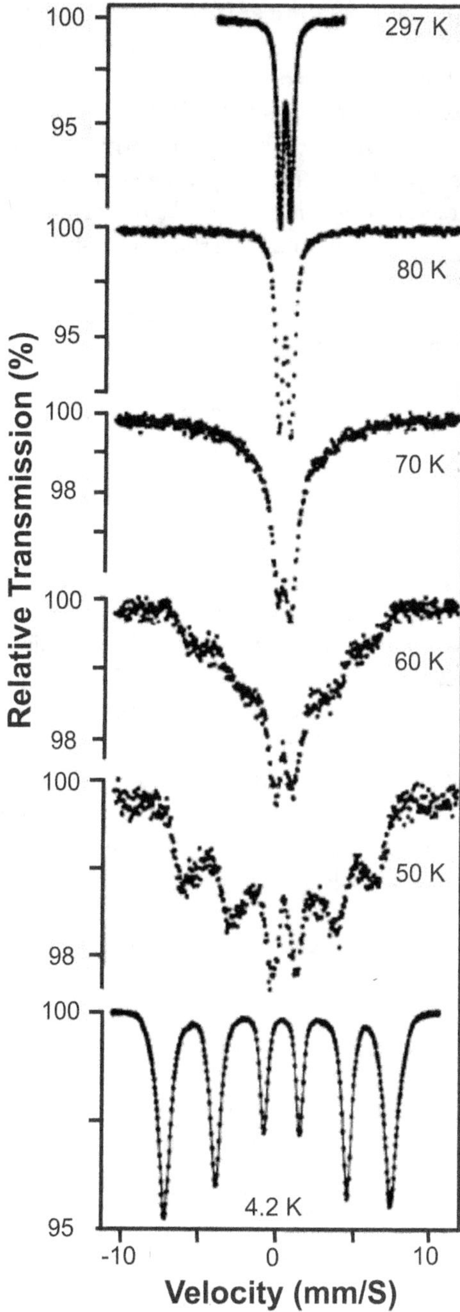

FIGURE 4.10 Mössbauer spectra taken at different temperatures for schwertmannites (ref. Bigham et al., 1990). Note the spectral pattern changes as temperature varies.

The IS values at room temperature are generally around 0.39 and 0.37 mm s^{-1}, and the QS values are approximately 0.89 and 0.62 mm s^{-1} with an average of about 0.65 mm s^{-1} (Bigham et al, 1990, Cashion and Murad, 2012; Schoepfer et al., 2019). This value is quite similar to that of a well-crystallized 6-line ferrihydrite, indicating similarities in their iron coordination environments. As temperature drops, magnetic ordering increases, but without clear sextets until around 78 K. At this temperature, evidence of magnetic ordering appears, with a hyperfine field of maximum probability of 34.6 T (Cashion and Murad, 2012).

By 4.2 K, the schwertmannite spectrum shows a fully magnetically ordered structure, with well-developed sextets characterized by IS of 0.47 and 0.48 mms^{-1} and QS of –0.11 and –0.08 mm s^{-1} (Bigham et al., 1990). This change in magnetic ordering, from doublets at room temperature to sextets at 4.2 K, indicates that schwertmannite undergoes a transition in its magnetic properties with decreasing temperature. The magnetic ordering temperature for schwertmannite is around 109 K, close to ferrihydrite's ordering temperature of 115 K, although the hyperfine fields are slightly lower due to the structural presence of SO_4^{2-} (Dakos et al., 2012).

Studies from a polish mine site suggests that As(V) ions influence the magnetic properties and temperature of schwertmannite's magnetic rearrangement and prevent splitting into sextets at 77 K for arsenic-rich samples (Parafiniuk and Siuda, 2006). Carlson et al. (2002) found that increasing the As/(As + S) ratio in synthetic schwertmannite precipitates increases QS, with values levelling off at higher As content (Figure 4.11). The highest QS values observed in As-rich schwertmannite are much larger than those of pure schwertmannite or ferrihydrite, indicating a distortion in Fe(O,OH) octahedra due to interaction with arsenate. As the temperature drops to 4.2 K, these samples exhibit broadened asymmetric sextets, with a magnetic hyperfine field lower than that of pure schwertmannite, suggesting reduced super exchange interactions between Fe ions due to As incorporation.

Mössbauer spectra of schwertmannite with trace contaminants like Cu(II), Sb(V), and P show similar IS and QS values to those of schwertmannite, with QS around 0.64–0.88 mm s^{-1} at 25°C (Vithana et al., 2018). In the presence of PO_4^{3-}, large QS occurred at low temperatures (QS ~ 0.7 mm s^{-1}), with an asymmetric doublet at 77 K, while goethite and lepidocrocite exhibit distinct sextets and sharp doublets, respectively (Schoepfer et al., 2019).

4.15 X-ray photoelectron spectroscopical patterns

X-ray photoelectron spectroscopy (XPS) is a widely used technique to study the surface composition of schwertmannite, especially for understanding

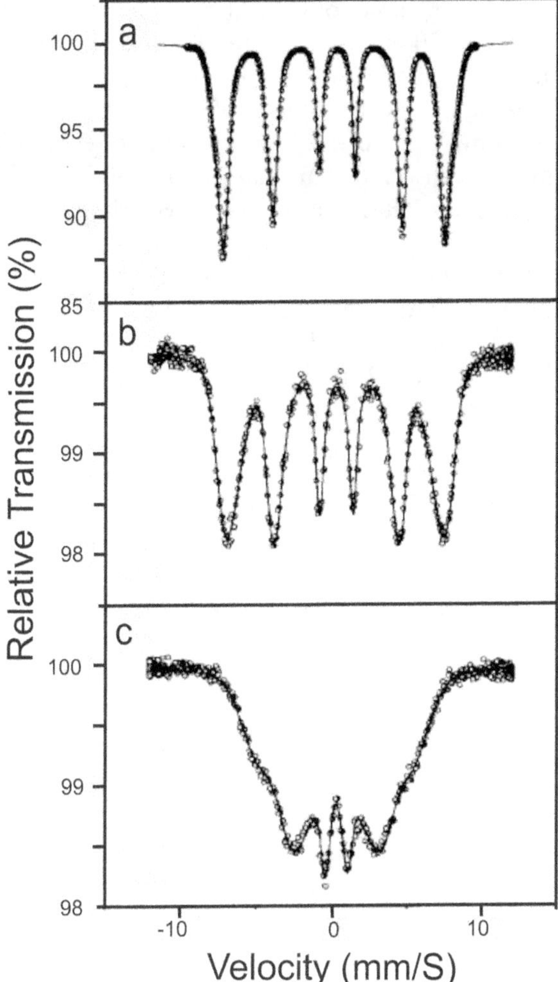

FIGURE 4.11 Mössbauer spectra taken at 4.2 K for (a) synthetic schwertmannite, (b) synthetic iron hydroxyarsenate (FeOHAs) with As/(As + S) = 1.0, and (c) synthetic iron arsenate with As/Fe = 1.0 (ref. Carlson et al., 2002).

the distribution of iron, sulphur, and oxygen, and to evaluate surface contaminant abundance. The main iron (Fe) signal comes from the Fe 2p region, which provides crucial information about the oxidation state of iron on schwertmannite surfaces. The principal Fe $2p_{3/2}$ peak typically occurs at binding energies (BE) of 711.7 or 710.8 eV, which are characteristic of Fe(III), along with a relatively intense satellite feature at around 720.0 eV

BE, also indicating the presence of Fe(III) (Fan et al., 2019b). Additionally, the Fe $2p_{1/2}$ peak of Fe(III) appears at a BE of 724.8 eV (Figure 4.12).

In contrast, the satellite feature at approximately 715 eV BE is attributed to Fe(II), providing evidence of its presence along with Fe(III) (Cutting et al., 2012). Deconvolution of the XPS spectra also reveals distinct peaks for Fe(II), commonly observed at around 710 and 707 eV BE, helping differentiate the oxidation states of iron on schwertmannite surfaces (Zhang et al., 2021).

This technique is also extensively used to study the structural changes in schwertmannite after As(V) adsorption, particularly for samples synthesized by rapid Fe^{2+} oxidation methods involving H_2O_2 and $KMnO_4$ (Cao et al., 2021; Zhang et al., 2021). The O 1s XPS peak provides insights into the bonding environment, with the peaks at 530.0 and 531.7 eV corresponding to Fe–O and Fe–OH bonds, respectively. Additionally, the peak at 532.2 eV is attributed to sulphate within the schwertmannite structure.

Upon As(V) adsorption at pH 7, the Fe–O peak area ratio for H_2O_2-mediated schwertmannite decreases from 25% to 9%, indicating a

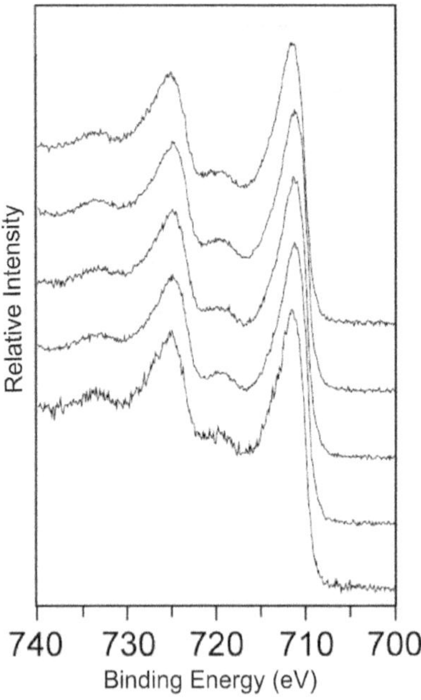

FIGURE 4.12 High-resolution X-ray photoelectron spectra showing the Fe 2p features of schwertmannite (ref. Cutting et al., 2012). The spectra represent schwertmannites synthesized through dialysis method containing 0, 0.3, 0.79, 2.3, and 4.13 wt% As(V). Note the Fe $2p_{3/2}$ peak at 711.7 or 710.8 eV and Fe $2p_{1/2}$ peak at 724.8 eV representing Fe(III).

significant structural alteration. In contrast, the $KMnO_4$-mediated schwertmannite shows no change in this ratio, suggesting that its mineral structure remains more stable under similar conditions. This implies that the H_2O_2-mediated schwertmannite undergoes more significant structural degradation due to As(V) incorporation than the $KMnO_4$-mediated one.

The S 2p spectra for schwertmannite show peaks at 168.5 and 169.8 eV, which correspond to surface-adsorbed SO_4^{2-} and structural SO_4^{2-} (Fe–SO_4). After As(V) adsorption, the Fe–SO_4 peak area decreases from 61% to 34% for the H_2O_2-mediated schwertmannite, indicating a considerable loss of structural SO_4^{2-}. This loss destabilizes the schwertmannite structure, suggesting potential transformation into ferrihydrite (Mazzetti and Thistlethwaite, 2002). In contrast, the $KMnO_4$-mediated

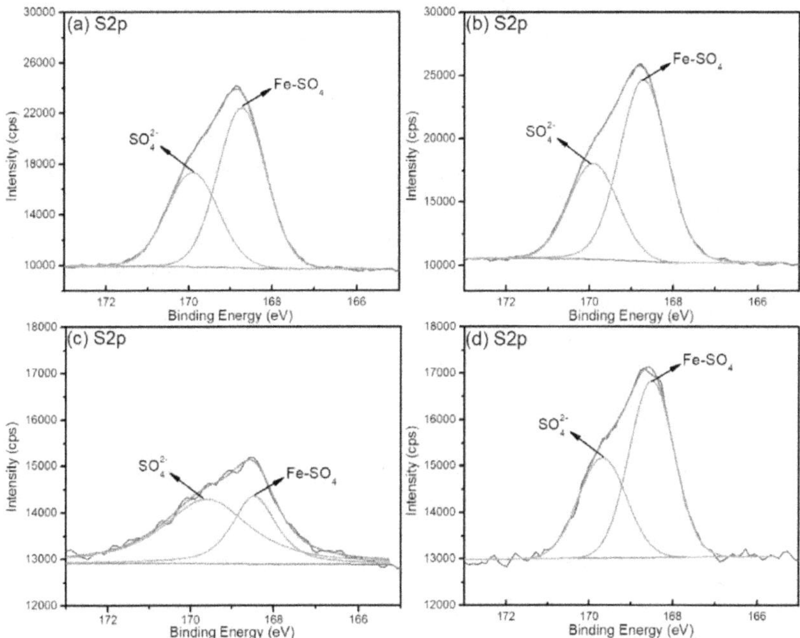

FIGURE 4.13 XPS spectra of S2p of schwertmannites synthesized by oxidation of Fe^{2+} with H_2O_2 (left side) or $KMnO_4$ (right side) method (ref. Cao et al., 2021). (a) and (b) represent S2p spectra before As(V) adsorption and (c) and (d) show corresponding spectra after As(V) adsorption at pH 7. Peaks at 168.5 and 169.8 eV correspond to surface-adsorbed SO_4^{2-} and structural SO_4^{2-} (Fe–SO_4). Note that the Fe–SO_4 peak area decreases from 61% to 34% for the H_2O_2-mediated schwertmannite, while there is a minor change from 65% to 62% for $KMnO_4$-mediated schwertmannite after As(V) adsorption.

schwertmannite shows minimal change in its SO_4^{2-}-content, with the Fe–SO_4 peak area decreasing only slightly, from 65% to 62%, confirming its structural stability (Figure 4.13).

Additional XPS peaks help to identify other S and As species. Sulphur species appear at 164 eV (S^0) and 167.65, 161.7, and 161.8 eV (S^{2-}). Arsenic detected at 45.4 and 44.6 eV are assigned to As(V), and 42.8 eV to arsenic sulphide (As_2S_3) (Zhang et al., 2021).

For Cr(VI)-containing schwertmannite, Cr 2p XPS peaks appear at 587, 579.4, 577.5, and 576.3 eV, representing Cr(VI), CrO_4^{2-}, Fe–CrO_4, and Cr–O, respectively. After 30 days of exposure to 4 mM Fe(II) in an anoxic medium, these peaks shift to 587.3, 586.4, 577.7, 576.8, and 576 eV, indicating partial reduction of Cr(VI) to Cr(III) (Fan et al., 2019a).

The XPS analysis of schwertmannite samples from the Kristineberg Zn–Cu mine, northern Sweden (Jönsson et al., 2005), revealed the elemental composition and oxidation states of key components. Iron constituted about 21% of the sample, mostly present as Fe(III) but with residuals of Fe(II). Sulphur accounted for approximately 16% in the form of sulphate (SO_4^{2-}). Oxygen made up about 59%, which was distributed as hydroxyl groups (26%), sulphate (16%), oxygen (10%), and water molecules (7%). Additionally, the presence of carbon in the range of 5%–14% was observed. This carbon content was attributed to the adsorption of hydrocarbons from the atmosphere and the contribution of organic material, such as fungi and bacteria, commonly found in environmental mineral samples.

4.16 Summary

Schwertmannite, both natural and synthetic, is widely characterized based on its colour, morphology, thermal behaviour, surface area, composition, and diffraction patterns. Various instrumental techniques such as XRD, SEM, HRTEM, ICP-MS, Raman and IR spectroscopy, Mössbauer spectroscopy, and XPS are used to study its physico-chemical properties.

Schwertmannite's colour ranges from reddish-yellow, yellowish-brown, to brownish shades, varying depending on its composition, origin, synthesis method, alteration, and impurity content. The Fe/S molar ratio of schwertmannite typically falls between 4.6 and 8.0, and its general formula is often expressed as $Fe_8O_8(OH)_{8-2x}(SO_4)_x{\cdot}nH_2O$, where $1 \leq x \leq 1.75$. However, the Fe/S ratio and values for "x" may significantly vary based on the presence of trace metals, synthesis pathways, and origin. Although data on schwertmannite density is limited, typical values range from 3.75 to 3.90 g cm^{-3}. The SSA of schwertmannite varies greatly with synthesis methods. Schwertmannite synthesized through bacterial oxidation of Fe^{2+}

and dialysis methods can have an SSA of up to 325 m^2 g^{-1}, while rapid oxidation of Fe^{2+} using H_2O_2 can result in SSA as low as 1.5 m^2 g^{-1}. Synthesis parameters like pH, temperature, ion availability, and the nature of the Fe(II) salt used can heavily impact SSA. The pH_{PZC} of schwertmannite ranges from 3.1 to 7.2, and its zeta potential and fractal geometry also vary significantly.

Schwertmannite typically exhibits a spheroidal morphology with acicular spines. Descriptions of its morphology include “pin-cushion”, “hedge-hog”, “filamentous”, and “ball-and-whisker”. The spines, or needles, usually range between 10–50 nm in length and 5–10 nm in width. The incorporation of contaminants or transformation processes often degrade these spines. Schwertmannite is characterized by eight diffraction peaks located at 0.489, 0.340, 0.256, 0.233, 0.197, 0.168, 0.151, and 0.146 nm, with the most intense peak at 0.256 nm. Aging, transformation, or contaminant uptake can shift the positions of these peaks.

Schwertmannite exhibits both endothermic and exothermic reactions upon heating. The first endothermic reaction occurs between 100°C and 300°C, attributed to the vaporization of sorbed H_2O and structural OH/H_2O, resulting up to 25% weight loss. This is followed by an exothermic reaction between 540°C and 580°C, which leads to the crystallization of $Fe_2(SO_4)_3$. A second endothermic reaction occurs between 650°C and 710°C, releasing SO_3 and further contributing to weight loss.

Spectral IR absorption bands for schwertmannite include those for $v_1(SO_4)$, $v_4(SO_4)$, and $v_3(SO_4)$ at 979–988, 1053–1196, and 1053–1196 cm^{-1}, respectively. The Fe–O stretch is observed at 667–700 and 408–460 cm^{-1}. Contaminants, such As(V), show bands at 820–828 cm^{-1} ($v_3(AsO_4^{3-})$). Mössbauer studies reveal IS and QS values around 0.39 and 0.89 mm s^{-1}, respectively. At low temperatures, schwertmannite displays magnetic ordering, with the magnetic field increasing as temperature decreases. The principal Fe $2p_{3/2}$ XPS peak for Fe(III) is observed at BE of 711.7 or 710.8 eV, with an intense satellite at around 720.0 eV BE. The Fe $2p_{1/2}$ peak occurs at 724.8 eV. The S2p spectra show peaks at 168.5 and 169.8 eV, corresponding to surface-adsorbed sulphate and structural sulphate, respectively.

References

Acero P, Ayora C, Torrentó C and Nieto J (2006). The behaviour of trace elements during schwertmannite precipitation and subsequent transformation into goethite and jarosite. *Geochimica et Cosmochimica Acta* 70: 4130–4139.

Acero P, Torrentó C and Ayora C (2005). Effect of schwertmannite ageing on acid rock drainage geochemistry. In *9th International Mine Water Congress*, IMWA Springer-Verlag, Spain, pp. 67–73.

Antelo J, Fiol S, Gondar D, López R and Arce F (2012). Comparison of arsenate, chromate and molybdate binding on schwertmannite: Surface adsorption vs anion-exchange. *Journal of Colloid and Interface Science* 386: 338–343.

Antelo J, Fiol S, Gondar D, Pérez C, López R and Arce F (2013). Cu(II) incorporation to schwertmannite: Effect on stability and reactivity under AMD conditions. *Geochimica et Cosmochimica Acta* 119: 149–163.

As K, Peiffer S, Uhuegbue PO, Joshi P, Kappler A, Marouane B and Hockmann K (2024). Sulphate affinity controls phosphate sorption and the proto-transformation of schwertmannite. *Chemical Geology* 653: 122043.

Asta MP, Cama J, Gault AG, Charnock JM and Queralt I (2007). Characterization of AMD sediments in the discharge of the Tinto Santa Rosa mine (Iberian pyritic belt, SW Spain). In Cidu R and Frau F (Eds), *Water in mining environments*, IMWA, Springer, pp. 1–5.

Bai S, Xu Z, Wang M, Liao M, Liang J, Zheng C and Zhou L (2012). Both initial concentrations of Fe(II) and monovalent cations jointly determine the formation of biogenic iron hydroxysulfate precipitates in acidic sulphate-rich environments. *Materials Science and Engineering C* 32: 2323–2329.

Baleeiro A, Fiol S, Otero-Fariña A and Antelo J (2018). Surface chemistry of iron oxides formed by neutralization of acidic mine waters: Removal of trace metals. *Applied Geochemistry* 89: 129–137.

Bigham JM, Algur ÖF, Jones FS and Tuovinen OH (2013). Solid-phase controls on lead partitioning in laboratory bioleaching solutions. *Hydrometallurgy* 136: 27–30.

Bigham JM, Carlson L and Murad E (1994). Schwertmannite, A new iron oxyhydroxysulphate from Pyhäsalmi, Finland, and other localities. *Mineralogical Magazine* 58: 641–648.

Bigham JM and Nordstrom DK (2000). Iron and aluminum hydroxysulfates from acid sulphate waters. In: Alpers CN, Jambor JL and Nordstrom DK (Eds), *Reviews in mineralogy and geochemistry, sulphate minerals: Crystallography, geochemistry, and environmental significance*, vol. 40. The Mineralogical Society of America, pp. 351–403.

Bigham JM, Schwertmann U, Carlson L and Murad E (1990). A poorly crystallized oxyhydroxysulfate of iron formed by bacterial oxidation of Fe(II) in acid mine waters. *Geochimica et Cosmochimica Acta* 54: 2743–2758.

Bigham JM, Schwertmann U, Traina SJ, Winland RL and Wolf M (1996). Schwertmannite and the chemical modeling of iron in acid sulphate waters. *Geochimica et Cosmochimica Acta* 60: 2111–2121.

Boily J, Gassman PL, Peretyazhko T, Szanyi J and Zachara JM (2010). FTIR spectral components of schwertmannite. *Environmental Science & Technology* 44: 1185–1190.

Boukhalfa C and Chaguer M (2012). Characterisation of sediments polluted by acid mine drainage in the northeast of Algeria. *International Journal of Sediment Research* 27: 402–407.

Brady KS, Bigham JM, Jaynes WF and Logan TJ (1986). Influence of sulphate on Fe-oxide formation: Comparisons with a stream receiving acid mine drainage. *Clays and Clay Minerals* 34: 266–274.

Burgos WD, Borch T, Troyer LD, Luan F, Larson LN, Brown JF, Lambson J and Shimizu M (2012). Schwertmannite and Fe oxides formed by biological low-pH

Fe(II) oxidation versus abiotic neutralization: Impact on trace metal sequestration. *Geochimica et Cosmochimica Acta* 76: 29–44.

Burton ED, Bush RT, Johnston SG, Watling KM, Hocking RK, Sullivan LA and Parker GK (2009). Sorption of arsenic(V) and arsenic(III) to schwertmannite. *Environmental Science & Technology* 43: 9202–9207.

Burton ED, Bush RT and Sullivan LA (2006). Sedimentary iron geochemistry in acidic waterways associated with coastal lowland acid sulphate soils. *Geochimica et Cosmochimica Acta* 70: 5455–5468.

Burton ED, Bush RT, Sullivan LA, Johnston SG and Hocking RK (2008). Mobility of arsenic and selected metals during re-flooding of iron- and organic-rich acid-sulphate soil. *Chemical Geology* 253: 64–73.

Burton ED, Bush RT, Sullivan LA and Mitchell DRG (2007). Reductive transformation of iron and sulphur in schwertmannite-rich accumulations associated with acidified coastal lowlands. *Geochimica et Cosmochimica Acta* 71: 4456–4473.

Burton ED, Johnston SG, Watling K, Bush RT, Keene AF and Sullivan LA (2010). Arseniceffects and behaviour in association with the Fe(II)-catalyzed transformation of schwertmannite. *Environmental Science & Technology* 44: 2016–2021.

Cao Q, Chen C, Li K, Sun T, Shen Z and Jia J (2021). Arsenic(V) removal behaviour of schwertmannite synthesized by $KMnO_4$ rapid oxidation with high adsorption capacity and Fe utilization. *Chemosphere* 264: 128398.

Caraballo MA, Rimstidt JD, Macías F, Nieto JM and Hochella Jr. MF (2013). Metastability, nanocrystallinity and pseudo-solid solution effects on the understanding of schwertmannite solubility.*Chemical Geology* 360–361: 22–31.

Carlson C, Bigham JM, Schwertmann U, Kyek A and Wagner F (2002). Scavenging of As from acid mine drainage by schwertmannite and ferrihydrite: A comparison with synthetic analogues. *Environmental Science &Technology* 36: 1712–1719.

Carrero S, Fernandez-Martinez A, Pérez-López R, Poulain A, Salas-Colera E and Nieto JM (2017). Arsenate and selenate scavenging by basaluminite: Insights into the reactivity of aluminum phases in acid mine drainage. *Environmental Science & Technology* 51: 28–37.

Cashion JD and Murad E (2012). Mössbauer spectra of the acid mine drainage mineral schwertmannite from the Sokolov Basin, Czech Republic. In: Stewart G (Ed), *36th annual condensed matter and materials meeting*, 31 January–3 February 2012. Wagga Wagga. pp. TP30:1–TP30:4.

Chen Q, Cohen DR, Andersen MS, Robertson AM and Jones DR (2022). Stability and trace element composition of natural schwertmannite precipitated from acid mine drainage. *Applied Geochemistry* 143: 105370.

Childs CW, Inoue K and Mizota C (1998). Natural and anthropogenic schwertmannites from Towada-Hachimantai National Park, Honshu, Japan. *Chemical Geology* 144: 81–86.

Choppala G and Burton ED (2018). Chromium(III) substitution inhibits the Fe(II)-accelerated transformation of schwertmannite. *Plos One* 13: e0208355.

Cruz-Hernández P, Carrero S, Pérez-López R, Fernandez-Martinez A, Lindsay MBJ, Dejoie C and Nieto JM (2019). Influence of As(V) on precipitation and transformation of schwertmannite in acid mine drainage-impacted waters. *European Journal of Mineralogy* 31: 237–245.

Cutting RS, Coker VS, Fellowes JW, Lloyd JR and Vaughan DJ (2009). Mineralogical and morphological constraints on the reduction of Fe(III) minerals by *Geobactersulfurreducens. Geochimica et Cosmochimica Acta* 73: 4004–4022.

Cutting RS, Coker VS, Telling ND, Kimber RL, Laan G, Pattrick RAD, Vaughan DJ, Arenholz E and Lloyd JR (2012). Microbial reduction of arsenic-doped schwertmannite by *Geobacter sulfurreducens. Environmental Science & Technology* 46: 12591–12599.

Dakos Z, Kupka D, Kovařik M, Jablonovská K, Krištúfek V and Achimovičová M (2012). Secondary iron minerals present in amd sediments from smolník abandoned mine. *Nova Biotechnologicaet Chimica* 11–2: 87–92.

Davidson LE, Shaw S and Benning LG (2008). The kinetics and mechanism of schwertmannite transformation to goethite and hematite under alkaline conditions. *American Mineralogist* 93: 1326–1337.

Dinelli E and Tateo F (2002). Different types of fine-grained sediments associated with acid mine drainage in the Libiola Fe–Cu mine area (Ligurian Apennines, Italy). *Applied Geochemistry* 17: 1081–1092.

Dold B and Fontboté L (2001). Element cycling and secondary mineralogy in porphyry copper tailings as a function of climate, primary mineralogy, and mineral processing. *Journal of Geochemical Exploration* 74: 3–55.

Dold B, Gonzalez-Toril E, Aguilera A, Lopez-Pamo E, Cisternas ME, Bucchi F and Amils R (2013). Acid rock drainage and rock weathering in Antarctica: Important sources for iron cycling in the southern ocean. *Environmental Science & Technology* 47: 6129–6136.

Dong S, Dou X, Mohan D, Pittman CU Jr. and Luo J (2015). Synthesis of graphene oxide/schwertmannite nanocomposites and their application in Sb(V) adsorption from water. *Chemical Engineering Journal* 270: 205–214.

Dou X, Mohan D and Pittman Jr. CU (2013). Arsenate adsorption on three types of granular schwertmannite. *Water Research* 47: 2938–2948.

Duan H, Liu Y, Yin X, Bai J and Qi J (2016). Degradation of nitrobenzene by Fenton-like reaction in a H_2O_2/schwertmannite system. *Chemical Engineering Journal* 283: 873–879.

Eskandarpour A, Onyango MS, Ochieng A and Asai S (2008). Removal of fluoride ions from aqueous solution at low pH using schwertmannite. *Journal of Hazardous Materials* 152: 571–579.

Fan C, Guo C, Chen M, Huang W, Wan J, Reinfelder JR and Li X, Zeng Y, Lu G and Dang Z (2019b). Transformation of cadmium-associated schwertmannite and subsequent element repartitioning behaviours. *Environmental Science and Pollution Research* 26: 617–627.

Fan C, Guo C, Zeng Y, Tu Z, Ji Y, Reinfelder JR, Chen M, Huang W, Lu G, Yi X and Dang Z (2019a). The behaviour of chromium and arsenic associated with redox transformation of schwertmannite in AMD environment. *Chemosphere* 222, 945–953.

Fernandez-Martinez A, Timon V, Roman-Ross G, Cuello GJ, Daniels JE and Ayora C (2010). The structure of schwertmannite, a nanocrystalline iron oxyhydroxysulfate. *American Mineralogist* 95: 1312–1322.

French RA, Caraballo MA, Kim B, Rimstidt JD, Murayama M and Hochella Jr. MF (2012). The enigmatic iron oxyhydroxysulfatenanomineral schwertmannite: Morphology, structure, and composition. *American Mineralogist* 97: 1469–1482.

Gan M, Song Z, Jie S, Zhu J, Zhu Y and Liu X (2016). Biosynthesis of bifunctional iron oxyhydrosulfate by Acidithiobacillus ferroxidans and their application to coagulation and adsorption. *Materials Science and Engineering C* 59: 990–997.

Gan M, Sun S, Zheng Z, Tang H, Sheng J, Zhu J and Liu X (2015a). Adsorption of Cr(VI) and Cu(II) by $AlPO_4$ modified biosynthetic schwertmannite. *Applied Surface Science* 356: 986–997.

Gan M, Zheng Z, Sun S, Zhu J and Liu X (2015b). The influence of aluminum chloride on biosynthetic schwertmannite and Cu(II)/Cr(VI) adsorption. *RSC Advances* 5: 94500.

Goswami A and Purkait MK (2014). Removal of fluoride from drinking water using nanomagnetite aggregated schwertmannite. *Journal of Water Process Engineering* 1: 91–100.

Götz AJ (2011). On the characteristics and growth of biominerals. PhD Dissertation. Fakultät für Geowissenschaften, Ludwig-Maximilians-Universität München, p. 158.

Gramp JP, Jones FS, Bigham JM and Tuovinen OH (2008). Monovalent cation concentrations determine the types of Fe(III) hydroxysulfate precipitates formed in bioleach solutions. *Hydrometallurgy* 94: 29–33.

Hockridge JG, Jones F, Loan M and Richmond WR (2009). An electron microscopy study of the crystal growth of schwertmannite needles through oriented aggregation of goethite nanocrystals. *Journal of Crystal Growth* 311: 3876–3882.

HoungAloune S, Hiroyoshi N and Ito M (2015). Effect of Fe(II) and Cu(II) on the transformation of schwertmannite to goethite under acidic condition. *International Journal of Chemical Engineering and Applications* 6: 32–37.

HoungAloune S, Hiroyoshi N and Ito M (2017). Effect of reaction temperature on schwertmannite synthesis, from simulated copper heap leach solutions. *International Journal of Management Information Technology and Engineering* 5: 27–34.

HoungAloune S, Kawaai T, Hiroyoshi N and Ito M (2014). Study on schwertmannite production from copper heap leach solutions and its efficiency in arsenic removal from acidic sulphate solutions. *Hydrometallurgy* 147–148: 30–40.

Jambor JL, Pertsev NN and Roberts AC (1995). New mineral names. *American Mineralogist* 80: 845–850.

Jiang W and Chen C (2005). Formation and transformation of schwertmannite in acid-mine drainage deposits of the Chinkuashih mining area, northern Taiwan. *Goldschmidt Conference Abstracts* A774.

Johnston SG, Burton ED and Moon EM (2016). Arsenic mobilization is enhanced by thermal transformation of schwertmannite. *Environmental Science & Technology* 50: 8010–8019.

Jönsson J, Jönsson J and Lövgren L (2006). Precipitation of secondary Fe(III) minerals from acid mine drainage. *Applied Geochemistry* 21: 437–445.

Jönsson J, Persson P, Sjoberg S and Lovgren L (2005). Schwertmannite precipitated from acid mine drainage: Phase transformation, sulphate release and surface properties. *Applied Geochemistry* 20: 179–191.

Ke C, Deng Y, Zhang S, Ren M, Liu B, He J, Wu R, Dang Z and Guo C (2024). Sulphate availability drives the reductive transformation of schwertmannite by co-cultured iron- and sulphate-reducing bacteria. *Science of the Total Environment* 906: 167690.

Khamphila K, Kodama R, Sato T and Otake T (2017). Adsorption and post adsorption behaviour of schwertmannite with various oxyanions. *Journal of Minerals and Materials Charaction and Engineering* 5: 90–106.

Kim JJ, Kim SJ and Tazaki K (2002). Mineralogical characterization of microbial ferrihydrite and schwertmannite, and non-biogenic Al-sulphate precipitates from acidmine drainage in the Donghae mine area, Korea. *Environmental Geology* 42: 19–31.

Kim H and Kim Y (2021). Schwertmannite transformation to goethite and the related mobility of trace metals in acid mine drainage. *Chemosphere* 269: 128.720.

Kumpulainen S, Carlson L and Raisanen ML (2007). Seasonal variations of ochreous precipitates in mine effluents in Finland. *Applied Geochemistry* 22: 760–777.

Kumpulainen S, Räisänen ML, Von der Kammer F and Hofmann T (2008). Ageing of synthetic and natural schwertmannites at pH 2–8.*Clay Minerals* 43: 437–448.

Kupka D, Pállová Z, Horňáková A, Achimovičová M and Kavečanský V (2012). Effluent water quality and the ochre deposit characteristics of the abandoned Smolník mine, East Slovakia. *Acta Montanistica Slovaca* 17: 56–64.

Lazaroff N, Sigal W and Wasserman A (1982). Iron oxidation and precipitation of ferric hydroxysulfates by resting *Thiobacillus ferrooxidans* cells. *Applied Environmental Microbiology* 43: 924–938.

Lee J and Kim Y (2008). A quantitative estimation of the factors affecting pH changes using simple geochemical data from acid mine drainage. *Environmental Geology* 55: 65–75.

Li ZY, Liang JR, Bai SY and Zhou LX. (2011). Characterization and As(III) adsorption properties of schwertmannite synthesized by chemical or biological procedures. *Acta Scientiae Circumstantiae* 31: 460–467.

Liao Y, Liang J and Zhou L (2011). Adsorptive removal of As(III) by biogenic schwertmannite from simulated As-contaminated groundwater. *Chemosphere* 83: 295–301.

Liu FW, Zhou J, Zhou LX, Zhang SS, Liu LL and Wang M (2015). Effect of neutralized solid waste generated in lime neutralization on the ferrous ion bio-oxidation process during acid mine drainage treatment. *Journal of Hazardous Materials* 299, 404–411.

Loan M, Hart RD, Cowley JM and Parkinson GM (2004). Evidence on the structure of synthetic schwertmannite. *American Mineralogist* 89: 1735–1742.

Loan M, Richmond WR and Parkinson GM (2005). On the crystal growth of nanoscale schwertamannite. *Journal of Crystal Growth* 275: e1875–e1881.

Mačingová E and Luptáková A (2014). Recovery of iron from acid mine drainage in the form of oxides. *Journal of the Polish Mineral Engineering Society* 15: 193–198.

Maillot F, Morin G, Juillot F, Bruneel O, Casiot C, Ona-Nguema G, Wang Y, Lebrun S, Aubry E, Vlaic G and Brown Jr. GE (2013). Structure and reactivity of As(III)- and As(V)-richschwertmannites and amorphous ferric arsenate sulfate from the Carnoule's acid mine drainage, France: Comparison with biotic and abiotic model compoundsand implications for As remediation. *Geochimica et Cosmochimica Acta* 104: 310–329.

Marescotti P, Carbone C, Comodi P, Frondini F and Lucchetti G (2012). Mineralogical and chemical evolution of ochreous precipitates from the Libiola Fe–Cu-sulfide mine (Eastern Liguria, Italy). *Applied Geochemistry* 27: 577–589.

Mazzetti L and Thistlethwaite PJ (2002). Raman spectra and thermal transformations of ferrihydrite and schwertmannite. *Journal of Raman Spectroscopy* 33: 104–111.

Miot J, Lu S, Morin G, Adra A, Benzerara K and Küsel K (2016). Iron mineralogy across the oxycline of a lignite mine lake. *Chemical Geology* 434: 28–42.

Mori JF, Lu S, Handel M, Totsche KU, Neu TR, Iancu VV, Tarcea N, Popp J and Küsel K (2016). Schwertmannite formation at cell junctions by a new filament-forming Fe(II)-oxidizing isolate affiliated with the novel genus Acidithrix. *Microbiology* 162: 62–71.

Murad E and Rojik P (2003). Iron-rich precipitates in a mine drainage environment: Influence of pH on mineralogy. *American Mineralogist* 88: 1915–1918.

Paikaray S and Peiffer S (2010). Dissolution kinetics of sulphate from schwertmannite under variable pH conditions. *Mine Water and the Environment* 29: 263–269.

Paikaray S and Peiffer S (2012). Biotic and abiotic schwertmannites as scavengers for As(III): Mechanisms and effects. *Water Air and Soil Pollution* 223: 2933–2942.

Paikaray S and Peiffer S (2015). Lepidocrocite formation kinetics from schwertmannite in Fe(II)-rich anoxic alkaline medium. *Mine Water and the Environment* 34: 213–222.

Paikaray S, Göttlicher J and Peiffer S (2011). Removal of As(III) from acidic waters using schwertmannite: Surface speciation and effect of synthesis pathway. *Chemical Geology* 283: 134–142.

Paikaray S, Schröder C and Peiffer S (2017). Schwertmannite stability in anoxic Fe(II)-rich aqueous solution. *Geochimica et Cosmochimica Acta* 217: 292–305.

Pállová Z, Kupka D and Achimovičová M (2010). Metal mobilization from AMD sediments in connection with bacterial iron reduction. *Mineralia Slovaca* 42: 343–347.

Parafiniuk J and Siuda R (2006). Schwertmannite precipitated from acid mine drainage in the Western Sudetes (SW Poland) and its arsenate sorption capacity. *Geological Quarterly* 50: 475–486.

Park JH, Han Y and Ahn JS (2016). Comparison of arsenic co-precipitation and adsorption by ironminerals and the mechanism of arsenic natural attenuation in a minestream. *Water Research* 106: 295–303.

Parviainen A, Cruz-Hernández P, Pérez-López R, Nieto JM and Delgado-López JM (2015). Raman identification of Fe precipitates and evaluation of As fate during phase transformation in Tinto and Odiel River Basins. *Chemical Geology* 398: 22–31.

Peiffer S and Gade W (2007). Reactivity of ferric oxides toward H_2S at Low pH. *Environmental Science & Technology* 41: 3159–3164.

Peretyazhko T, Zachara JM, Boily JF, Xia Y, Gassman PL, Arey BW and Burgos WD (2009). Mineralogical transformations controlling acid mine drainage chemistry. *Chemical Geology* 262: 169–178.

Qiao X, Liu L, Shi J, Zhou L, Guo Y, Ge Y, Fan W, Liu F (2017). Heating changes bio-schwertmannite microstructure and arsenic(III) removal efficiency. *Minerals* 7: 1–14.

Regenspurg S, Brand A and Peiffer S (2004). Formation and stability of schwertmannite in acidic mining lakes. *Geochimica et Cosmochimica Acta* 68: 1185–1197.

Reichelt L and Bertau M (2015). Transformation of nanostructured schwertmannite and 2-Line-ferrihydrite into hematite. *Journal of Inorganic and General Chemistry* 641: 1696–1700.

Santofimia E, López-Pamo E and Montero E (2015). Selective precipitation of schwertmannite in a stratified acidic pit lake of Iberian Pyrite Belt. *Mineralogical Magazine* 79: 497–513.

Scheinost AC and Schwertmann U (1999). Color identification of iron oxides and hydroxysulfates: Use and limitations. *Soil Science Society of America Journal* 63: 1463–1471.

Schoepfer VA and Burton ED (2021). Schwertmannite: A review of its occurrence, formation, structure, stability and interactions with oxyanions. *Earth-Science Reviews* 221: 103811.

Schoepfer VA, Burton ED, Johnston SG and Kraal P (2017). Phosphate-imposed constraints on schwertmannite stability under reducing conditions. *Environmental Science & Techno*logy 51: 9739–9746.

Schoepfer VA, Burton ED and Johnston SG (2019). Contrasting effects of phosphate on the rapid transformation of schwertmannite to Fe(III) (oxy)hydroxides at near-neutral pH. *Geoderma* 340: 115–123.

Sidenko NV and Sherriff BL (2005). The attenuation of Ni, Zn and Cu, by secondary Fe phases of different crystallinity from surface and ground water of two sulfide mine tailings in Manitoba, Canada. *Applied Geochemistry* 20: 1180–1194.

Song J, Jia S-Y, Ren H-T, Wu S-H and Han X (2015). Application of a high-surface-area schwertmannite in the removal of arsenate and arsenite. *International Journal of Environmental Science &Technology* 12: 1559–1568.

Šubrt J, Michalková E, Boháček J, Lukáč J, Gánovská Z and Máša B (2011). Uniform particles formed by hydrolysis of acid mine drainage with urea. *Hydrometallurgy* 106: 12–18.

Tresintsi S, Simeonidis K, Pliatsikas N, Vourlias G, Patsalas P and Mitrakas M (2014). The role of SO_4^{2-} surface distribution in arsenic removal by iron oxy-hydroxides. *Journal of Solid State Chemistry* 213: 145–151.

Tresintsi S, Simeonidis K, Vourlias G, Stavropoulos G and Mitrakas M (2012). Kilogram-scale synthesis of iron oxy-hydroxides with improved arsenic removal capacity: Study of Fe(II) oxidation-precipitation parameters. *Water Research* 46: 4255–4267.

Vithana CL, Sullivan LA, Bush RT and Burton ED (2015). Schwertmannite in soil materials: Limits of detection of acidified ammonium oxalate method and differential X-ray diffraction. *Geoderma* 249–250: 51–60.

Vithana CL, Johnston SG and Dawson N (2018). Divergent repartitioning of copper, antimony and phosphorus following thermal transformation of schwertmannite and ferrihydrite. *Chemical Geology* 483: 530–543.

Wan J, Guo C, TuZ Zeng Y, Fan C, Lu G and Dang Z (2018). Microbial reduction of Cr(VI)-loaded schwertmannite by *Shewanellaoneidensis* MR-1. *Geomicrobiology Journal* 35: 727–734.

Wang H, Bigham JM and Tuovinen OH (2006). Formation of schwertmannite and its transformation to jarosite in the presence of acidophilic iron-oxidizing microorganisms. *Material Science and Engineering C* 26: 588–592.

Wang W, Song J and Han X (2013). Schwertmannite as a new Fenton-like catalyst in the oxidation ofphenol by H_2O_2. *Journal of Hazardous Materials* 262: 412–419.

Wang X, Gu C, Feng X and Zhu M (2015). Sulfate local coordination environment in schwertmannite. *Environmental Science & Technology* 49: 10440–10448.

Wang Y, Gao M, Huang W, Wang T and Liu Y (2020). Effects of extreme pH conditions on the stability of As(V)-bearing schwertmannite. *Chemosphere* 251: 126427.

Webster JG, Swedlund PJ and Webster KS (1998). Trace metal adsorption onto a acid mine drainage iron(III) oxy hydroxy sulphate. *Environmental Science & Technology* 32: 1361–1368.

Wu Y, Guo J, Jiang D, Zhou P, Lan Y and Zhou L (2012). Heterogeneous photocatalytic degradation of methyl orange in schwertmannite/oxalate suspension under UV irradiation. *Environmental Science and Pollution Research* 19: 2313–2320.

Xie Y, Lu G, Ye H, Yang C, Xia D, Yi X, Reinfelder J and Dang Z (2017). Fulvic acid induced the liberation of chromium from CrO_4^{2-} substituted schwertmannite. *Chemical Geology* 475: 52–61.

Xiong H, Lu X and Wang R (2020). Anionic effect on nanostructure and morphology of bio-schwertmannite dynamically produced within cellular reproduction. *Nanomaterials and Nanotechnology* 10: 1–10.

Xiong HX, Liao YH and Zhou LX (2008). Influence of chloride and sulphate on formation of akaganéite and schwertmannite through ferrous biooxidation by *Acidithiobacillusferrooxidans* cells. *Environmental Science & Technology* 42: 8681–8696.

Xiong H, Hu D, Shi K, Zhu S and Xu Y (2023). Adsorptive removal of arsenic ions from contaminated water using low-cost schwertmannites and akaganéites. *Materials Chemistry and Physics* 297: 127411.

Xu Y, Yang M, Yao T and Xiong H (2014). Isolation, identification and arsenic-resistance of Acidithiobacillus ferrooxidans HX3 producing schwertmannite. *Journal of Environmental Sciences* 26: 1463–1470.

Yan S, Zheng G, Meng X and Zhou L (2017). Assessment of catalytic activities of selected iron hydroxysulphates biosynthesized using Acidithio bacillus ferrooxidans for the degradation of phenol in heterogeneous Fenton-like reactions. *Separation and Purification Technology* 185: 83–93.

Ying H, Feng X, Zhu M, Lanson B, Liua F and Wang X (2020). Formation and transformation of schwertmannite through direct Fe^{3+} hydrolysis under various geochemical conditions. *Environmental Science Nano* 7: 2385.

Yu J, Heo B, Choi I, Cho J and Chang H (1999). Apparent solubilities of schwertmannite and ferrihydrite in natural stream waters polluted by mine drainage. *Geochimica et Cosmochimica Acta* 63: 3407–3416.

Zhang C, Zhang Z, Chen M and Fu D (2017). The influence of fractal nature on schwertmannite adsorption properties. *RSC Advances* 7: 27895–27899.

Zhang S, Jia S, Yu B, Liu Y, Wu S and Han X (2016). Sulfidization of As(V)-containing schwertmannite and its impact on arsenic mobilization. *Chemical Geology* 420: 270–279.

Zhang Z, Bi X, Li X, Zhao Q and Chen H (2018). Schwertmannite: Occurrence, properties, synthesis and application in environmental remediation. *RSC Advance* 8: 33583–33599.

Zhang Y, Gao K, Dang Z, Huang W, Reinfelder JR and Ren Y (2021). Microbial reduction of As(V)-loaded schwertmannite by *Desulfosporosinus meridiei*. *Science of the Total Environment* 764: 144279.

Zhang Z, Guo G, Li X, Zhao Q, Bi X, Wu K and Chen H (2019). Effects of hydrogen-peroxide supply rate on schwertmannite microstructure and chromium(VI) adsorption performance. *Journal of Hazardous Materials* 367: 520–528.

Zhu J, Chen F and Gan M. (2020). Controllable biosynthesis of nanoscale schwertmannite and the application in heavy metal effective removal. *Applied Surface Science* 529: 147012.

5

RETENTION OF CONTAMINANTS BY SCHWERTMANNITE

5.1 Introduction

It is well established that schwertmannite is able to entrap trace metal(oid)s, (oxo)anions, rare earth elements (REEs), and organic pollutants in large quantities. In acidic mine drainage–affected waters, four major mechanisms are discussed to be responsible for the natural attenuation of trace metals and metalloids (Chapman et al., 1983):

1 Precipitation of secondary phases, such as schwertmannite, followed by sedimentation
2 Sorption and co-precipitation of dissolved species (cations, anions, and trace metals) onto newly formed precipitates
3 Adsorption of dissolved species onto suspended materials such as clay particles
4 Dilution by tributaries and groundwater

Schwertmannite retains contaminants through different mechanisms depending on pH, surface charge, and the nature of the contaminant. Like other iron oxyhydroxides, schwertmannite has a positive surface charge at pH levels below its point of zero charge (pH_{PZC}) and a negative charge above this value. Surface-adsorbed sulphate ions (SO_4^{2-}) play a significant role in contaminant behaviour when interacting with schwertmannite. For example, surface SO_4^{2-} can enhance the sorption of trace metals through complexation under acidic conditions. Some contaminants, such as Cr(VI), may replace structural SO_4^{2-}, while others, like arsenic (As),

DOI: 10.1201/9781003583875-5

substitute surface-bound SO_4^{2-}. Co-precipitation is another significant retention mechanism shown to be effective in both abiotic and biotic schwertmannite formation. This process works well for pollutants like fluoride (F^-), which is retained effectively through co-precipitation (Zhu et al., 2020). Additionally, trace metals can substitute iron in schwertmannite when their ionic radii are similar. Retention of trace metals also impacts schwertmannite's structural properties, leading to lowered X-ray diffraction (XRD) peak intensities, weakened infrared (IR) peak intensities, and destruction of surface spines, among other changes.

The high affinity of schwertmannite for contaminants has also driven research in the application of this mineral as a sequestering agent to be used, e.g., as a filter material in the treatment of contaminated waters. In this chapter, we first present examples from field studies on the natural attenuation of trace metals and metalloids in acid mine drainage (AMD) environments and the involvement of schwertmannite. Due to the strong potential of schwertmannite to remove contaminants, systematic studies had been performed with several contaminants to understand their uptake patterns, which differ in their binding behaviour and their effects on the physico-chemical properties of schwertmannite. This subchapter is followed by a discussion about the mechanisms and controls of contaminant retention. Finally, we report about the application of schwertmannite as an adsorbent for the retention of non-AMD contaminants.

5.2 Natural attenuation of trace metal(loid)s in AMD-affected environments and the role of schwertmannite

Significant metal(loid) release occurs through chemical and microbial leaching of mine wastes, but natural attenuation mechanisms often reduce these concentrations as they move downstream from mine drainage areas. Three key processes contribute to this natural attenuation:

1 Adsorption: Precipitates formed from Fe- and SO_4^{2-}-rich mine drainage serve as primary adsorbing agents. Schwertmannite dominates in acidic conditions ($pH < 5$), while ferrihydrite is prevalent in near-neutral pH environments. In areas with high aluminium (Al^{3+}) content, aluminium oxyhydroxides also play a significant role in trapping contaminants (Yu and Heo, 2001).
2 Co-precipitation: Trace metals can become structurally incorporated into newly formed minerals during the precipitation process. Schwertmannite and ferrihydrite are the dominant phases during this process where trace metals become strongly bound within the structure of these secondary precipitates.

3 Dilution: The reduction in trace metal concentrations downstream is also attributed to the dilution of contaminated water with clean stream flows and rainwater. This is particularly notable in areas with heavy seasonal rainfall (e.g., Youngdong AMD, South Korea), where high rainfall reduces contaminant concentrations and raises pH during the rainy season (Yu and Heo, 2001). In contrast, lower rainfall seasons lead to higher contaminant concentrations and acidic conditions similar to the source areas.

These three processes often work together, making it difficult to distinguish their individual contributions. However, indicators such as the formation of new mineral phases, changes in pH, and seasonal variations in contaminant levels can provide clues as to which process might be dominating in a given context.

Natural As enrichment in schwertmannite has been observed to be particularly high around mine drainage sites, e.g., Radzimowice mine, Poland, where As concentrations reach 5.2 wt%. Schwertmannite's ability to retain As primarily occurs through adsorption, with about two-thirds of the As being adsorbed onto surface sites, while structural incorporation accounts for around 30%. Only about 1% of As is retained as ferric arsenate. Schwertmannite formations enriched in As have been reported to occur also in mine sites from the Czech Republic (2.38 wt% As_2O_5; Regenspurg and Peiffer, 2005) and from Japan (2.13 wt% As_2O_5; Fukushi et al., 2003a).

Research conducted in the Iberian Pyrite Belt at the Monte Romero mine (Spain) further highlights the selective affinity of schwertmannite for As and vanadium (V) compared to other metals like Zn, Co, Ni, Pb, and Cd (Acero et al., 2006). The binding of As and V occurs via aqueous anionic complexes ($H_2AsO_4^-$ and $VO_2SO_4^-$), while other metals are likely present as uncharged divalent complexes (MSO_4^0). Schwertmannite, with a positively charged surface under acidic conditions (pH ~ 3.0), readily adsorbs these anionic species. The high affinity of As and V for schwertmannite results in >99% retention in the solid phase, making them largely unavailable to the aqueous phase.

Similarly, Dakos et al. (2012) observed significant trace metal retention by schwertmannite precipitated from mine drainage at the abandoned Smolník mine, Slovakia, despite a low specific surface area (SSA; 4.54 $m^2 g^{-1}$). The precipitated schwertmannite contained exceptionally high concentrations of As, Mo, Mn, Zn, Ni, Cd, Pb, Cr, and Sb, with minimal dissolved metal concentrations remaining in the water.

Although natural attenuation processes significantly reduce trace metal concentrations through the precipitation of schwertmannite, uptake capacities of natural schwertmannite specimens are generally lower than those of schwertmannite synthesized in the laboratory. This difference can be attributed to the higher purity of laboratory-synthesized schwertmannite, whereas naturally precipitating schwertmannite often contains impurities and is associated with minor secondary phases. Naturally occurring schwertmannite frequently co-precipitates with Al^{3+} from mine effluents, either as an impurity or in the form of separate Al-rich phases, such as $Al(OH)_3$ or $Al_4(SO_4)(OH)_{10}·15H_2O$. These impurities and associated phases reduce the adsorption efficiency of naturally formed schwertmannite compared to its synthetic counterpart (Baleeiro et al., 2018).

Natural attenuation processes cause trace metals to be distributed differently across various sediment constituents in mine drainage environments.

Sidenko and Sherriff (2005) documented natural attenuation of Ni, Zn, and Cu via schwertmannite precipitation around the Ruttan and Thompson mines, Canada. Similarly, Webster et al. (1998) observed natural attenuation of Cu, Zn, Pb, and Cd near the Tui Pb–Zn mine, New Zealand, where schwertmannite was a key phase in metal attenuation, followed by goethite further downstream. Similar natural attenuation of heavy metals by schwertmannite has also been found around Dabaoshan mine, China (Zhao et al., 2012). Research performed on REE behaviour in relation to schwertmannite is limited, but it suggests that maximum uptake occurs at pH values between 4.5 and 6.5, depending on the REE type. Light REEs are less readily absorbed than heavy REEs for any given pH (Lozano et al., 2020). Since environmental threats posed by REEs are uncommon, the role of schwertmannite in REE retention is not covered in detail here.

The in-situ removal of heavy metals such as Zn, Cu, and Mn from AMD has been successfully carried out through actively precipitating schwertmannite in effluents of the Dexing Cu mine, China, with Fe and SO_4^{2-} contents of 1285 and 8011 mg L^{-1}, respectively, and a pH of 2.2 (Wang et al., 2019). To this end, Fe^{3+} was reduced to Fe^{2+} by using iron (Fe^0) powder. In a cyclic process, the produced Fe^{2+} was oxidized to Fe^{3+} by H_2O_2 facilitating the formation of a reddish-brown to orange precipitate schwertmannite. This step was followed by lime neutralization up to pH 7.5, which resulted in ~95% removal of all heavy metals. The amount of lime required was substantially reduced through this combined chemical mineralization and lime neutralization (1.2 mg CaO per litre wastewater compared to 3.3 mg L^{-1} when only lime was added). Sludge production was also lower (3.3 mg L^{-1} wastewater vs. 11.1 mg L^{-1}, respectively).

The neutralization of coal mine drainage in Pennsylvania, USA, similarly led to the retention of high concentrations of metals such as Ni, Co, Mn, and Zn onto solid phases during treatment. These solid phases were composed of mixtures of schwertmannite, ferrihydrite, and goethite (Burgos et al., 2012). The retention of these metals was enhanced when the neutralization pH exceeded 7.1, as the metals were retained through electrostatic attractions at pH > pH_{PZC} of the precipitating minerals. The process also effectively trapped high concentrations of Al^{3+}, which precipitated in the form of hydrobasaluminite ($Al_4(SO_4)(OH)_{10}\cdot 15H_2O$), amorphous $Al(OH)_3$, or co-precipitated with schwertmannite and ferrihydrite. The aluminium co-precipitation and the formation of these solid phases contributed significantly to the efficiency of the neutralization process, allowing for better control of heavy metal and aluminium concentrations in the treated drainage.

Amendments of As-contaminated paddy fields with a soil As content of 85.6 mg kg^{-1} with schwertmannite have been attempted by Wang et al. (2023), who applied biosynthesized schwertmannite together with phosphate fertilizer ($Ca(H_2PO_4)_2$, 0.75–600 mg kg^{-1}) for a 120-day period of rice cultivation. It was found that As release increased with time upon addition of phosphate fertilizer only, but decreased substantially in the presence of 0.5 wt% schwertmannite (855 vs. 395 µg L^{-1}). Arsenic enrichment in root and rice grains was reduced significantly by the amendment by only 0.5 wt% schwertmannite. The As content in rice grains was 1.06–1.47 mg kg^{-1} when treated with Pi fertilizer that reduced to 0.38–0.63 mg kg^{-1} when amended with 0.5 wt% schwertmannite.

Collectively these studies underpin the prominent role schwertmannite plays in the attenuation of contaminants in AMD-affected environments. Moreover, this research also demonstrates that schwertmannite can be even used as a reactant to treat contaminated waters. Yet, contaminants interact specifically with schwertmannite, which we will discuss in the following chapters.

5.3 Arsenic retention

Arsenic contamination of groundwater and surface water is a serious environmental threat in many countries. Both human activities and natural processes contribute to arsenic enrichment in the environment. Human sources include industrial waste from arsenic-containing products, agricultural activities, and pharmaceutical laboratories. Natural sources include the weathering of As-rich minerals and coal combustion by-products, such as arsenopyrite (FeAsS) and arsenical pyrite (FeS_2–As) (Carlson et al., 2002).

The chemistry of As in solution plays a critical role in determining its mobility in natural systems and its availability to plants, crops, and living organisms. Arsenic exists in four stable oxidation states: +5, +3, 0, and –3, with the +5 (arsenate) and +3 (arsenite) forms being the most common ones. These two forms can interconvert depending on the redox conditions of the environment. As a redox-sensitive contaminant, behaviour of As is dynamic and largely controlled by the geochemical and microbial conditions of the system, such as pH and redox potential. While arsenate (As(V)) is more stable under oxidizing conditions and arsenite (As(III)) under reducing conditions, both forms can coexist in the same environment due to slow redox reactions (Smedley and Kinniburgh, 2002). This influences its toxicity to humans, with As(III) being more toxic than As(V). Exposure to As can increase the risk of cancers in the liver, lungs, skin, bladder, and kidneys (Mandal and Suzuki, 2002; Duker et al., 2005; Sthiannopkao et al., 2010). In natural environments, As(V) primarily exists as $H_2AsO_4^-$ and $HAsO_4^{2-}$ with pK_a values of 2.1, 6.7, and 11.2, while As(III) exists mainly as H_3AsO_3 with pK_a values of 9.1, 12.2, and 13.4 (Mohan and Pittman, 2007). Arsenate is present as H_3AsO_4 and $H_2AsO_4^-$ at pH levels below 3.0, as $H_2AsO_4^-$ at pH 4.0–5.5, and as AsO_4^{3-} and $HAsO_4^{2-}$ at pH levels above 7.0. For As(III), H_3AsO_3 is the dominant species at pH below 7.0, and both H_3AsO_3 and $H_2AsO_3^-$ are present at pH above 7.0. Various methylated arsenic compounds, such as monomethyl arsenic acid (MMA), dimethyl arsenic acid (DMA), arsenobetaine (AsB), and trimethyl arsine oxide (TMAO), have been detected in marine and estuarine sediments (Reimer and Thompson, 1988). The formation of these organo-arsenic compounds is believed to involve biological processes, where inorganic As is converted into organic derivatives as a detoxification mechanism (Cullen and Reimer, 1989). In addition, the formation of thiolated arsenic species, in which oxygen is replaced by sulphur in both arsenates and arsenites, has been reported in sulfidic waters (Wilkin et al., 2003).

The ferrous iron (Fe(II)) released upon oxidative dissolution of sulphide-rich wastes around mine sites (see Chapter 1) becomes oxidized, either by bacteria like *Acidithiobacillus ferrooxidans* or abiotically by atmospheric oxygen, leading to the formation of Fe(III)-containing insoluble precipitates, including jarosite, schwertmannite, ferrihydrite, and goethite, as the pH shifts from acidic to alkaline (Bigham et al., 1996a, 1996b; Burgos et al., 2012). Arsenic naturally binds to these precipitates during or after their formation, limiting its mobility downstream. The effectiveness of these mineral phases in adsorbing As varies depending on their crystallinity. Schwertmannite, with its high surface area (up to 300 m^2 g^{-1}, please refer Chapter 4), has a strong capacity to retain As. Since As is sensitive to changes in oxidation state, its interaction with schwertmannite varies depending on the pH of the water and the presence of other ions.

Schwertmannite is particularly of interest due to its likely presence in areas impacted by AMD, where As is a predominant contaminant. Schwertmannite typically incorporates more As compared to other iron minerals, such as ferrihydrite, lepidocrocite, hematite, akaganéite, and goethite. This superior As uptake is due to its amorphous structure, high surface area, and exchangeable SO_4^{2-} content (Song et al., 2015; Park et al., 2016). This efficiency is consistent in both adsorption and co-precipitation processes. The retention of As on schwertmannite is influenced by several factors, including the distribution of As species (As(III)/As(V)) at a given pH, the surface charge of schwertmannite at that pH, and the sorption mechanism—whether it involves ligand exchange with schwertmannite-bound SO_4^{2-} or structural binding with Fe^{3+}. As(V) retention by schwertmannite is higher at acidic pH levels, while As(III) shows better retention at alkaline pH levels (Song et al., 2015).

Retention is controlled by the interaction between negatively charged $H_2AsO_4^-$ species at pH levels below 5.5 and the positively charged surface of schwertmannite when the pH is below its pH_{PZC} (Fan et al., 2019). Schwertmannite's pH_{PZC} is often found to be around 7.0 (Jönsson et al., 2005; Khamphila et al., 2017); so at pH levels below 7.0, the surface has a positive charge, while above 7.0, it becomes negatively charged. At higher pH levels, the negatively charged schwertmannite surface repels As(V) species (AsO_4^{3-} and $HAsO_4^{2-}$), reducing the retention capacity. However, schwertmannite still retains As(V) even at pH levels above 5.5 due to its strong affinity to Fe^{3+} (Jönsson et al., 2005).

Compared to other pollutants like phosphate, chromate, and selenate, schwertmannite has a higher retention capacity for As(V) in the order of 1.023, 0.934, 0.723, and 0.313 mmol g^{-1} for As(V), PO_4, Cr(VI), and Se(VI), respectively (Khamphila et al., 2017). The release of SO_4^{2-} from schwertmannite is directly proportional to the uptake of different oxyanions, with the highest SO_4^{2-} release observed during As(V) sorption, and the lowest during Se(VI) sorption. This is due to the exchange of SO_4^{2-} with the sorbed oxyanions (Figure 5.1a). The adsorption experiments started at a pH of 7 and dropped to values between 3.5 and 4 after 30 days of reaction, with lowest values obtained at highest initial anion concentration. A substantial release of iron (Fe) was observed after 30 days, the release rates showing an inverse trend compared to the release rates of SO_4^{2-} (Figure 5.1b). The least Fe is released in the As(V) retention experiments, and the most Fe is released during Se(VI) sorption (Figure 5.1b). The release rates of both SO_4^{2-} and Fe were found to be directly proportional to the initial concentrations of the solutes. Solid-phase analysis indicated that both As(V) and PO_4^{3-} stabilized schwertmannite, preventing any mineralogical changes even after 30 days of exposure at pH 7.0. In contrast, goethite formed in media containing Cr(VI), Se(VI), and SO_4^{2-} after 30, 14, and 7 days, respectively.

Compared to other iron minerals like goethite and hematite, arsenic retention by schwertmannite is generally strong enough to withstand

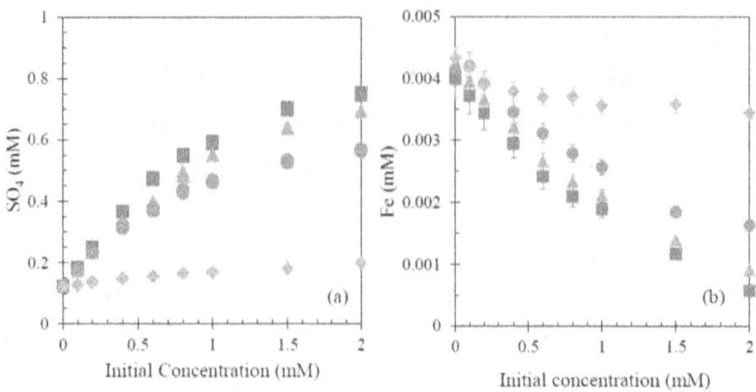

FIGURE 5.1 SO_4 (a) and Fe (b) concentrations released from schwertmannite after 30 days of reaction at a final pH between 3.5 and 4 as a function of the initial concentration of anions during the adsorption process (ref. Khamphila et al., 2017). (Square represents arsenate, triangle = phosphate, circle = chromate, and diamond = selenate).

release once becoming exposed to aqueous solution. This strong retention is primarily due to the structural binding of As, while surface-adsorbed As is more prone to mobilization. Arsenic retained through co-precipitation is more stable than when retained through adsorption (Alarcón et al., 2014; Wang et al., 2020, cf. below). A study by Wang et al. (2020) examined the synthesis of schwertmannite through rapid oxidation using H_2O_2 in an As(V)-rich medium, and the interaction of As(V) with pure schwertmannite prepared by the same method. The results showed higher As/Fe and As/S ratios of 0.07 and 0.32 in schwertmannite for co-precipitated arsenic, compared to 0.04 and 0.07 for adsorbed As. In contrast, surface-sensitive energy-dispersive X-ray spectroscopic mapping often shows higher As/Fe ratios for adsorbed arsenic compared to co-precipitated arsenic (Park et al., 2016), which implies a higher surface loading during adsorption processes than during co-precipitation.

Arsenic(V) is more strongly retained by iron oxyhydroxides compared to As(III) (Dixit and Hering, 2003; Egal et al., 2009). Low As(III) retention (20%–40%) has been observed at pH 3.0, compared to higher retention of As(V) (89%–98%) during the synthesis of schwertmannite from an As(III)-enriched medium (0.0–5.12 mM) (Maillot et al., 2013). Therefore, oxidation of As(III) to As(V) followed by adsorption onto suitable sorbents is commonly used to maximize As removal from contaminated water and soil. Oxidizing agents such as permanganate, chlorine, chloramines, and ozone are typically employed for this purpose. However, no significant oxidation of As(III) to As(V) occurs during aerobic retention studies (Liao et al., 2011; Paikaray et al., 2011, 2014).

Some studies have explored the conversion of As(III) to As(V) in the presence of As(III)-oxidizing bacteria, along with Fe(II)-oxidizing bacteria, to promote schwertmannite formation in acidic, Fe- and SO_4^{2-}-rich environments (Yang et al., 2017). As-contaminated soil was treated with the As(III)-oxidizing bacterial strain YZ-1. The oxidation of soil As(III) by the bacteria was enhanced by increasing the carbon sources (glucose, sucrose) and nitrogen sources (yeast, ammonium nitrate), with glucose and yeast providing the best results as carbon and nitrogen sources, respectively, at pH 8.0. The addition of schwertmannite to the contaminated soil, along with the bacterial strain YZ-1, further improved arsenic removal efficiency.

An analysis of the precipitates formed under varying As(III) and As(V) concentrations in the presence of ferrous sulphate shows that schwertmannite is the only mineral present at As(III)/Fe(III) ratios of 0.0–0.8. However, schwertmannite formation ceases at As(V)/Fe(III) ratios greater than 0.2, with As(V)–Fe(III) oxyhydroxysulphate or ferric arsenate appearing instead (Maillot et al., 2013). This phase shift from As(V)–schwertmannite to other arsenic-bearing minerals may be due to the excess As(V) loading, which displaces SO_4^{2-} in schwertmannite, causing an imbalance in the Fe(III)/As(V) molar ratio and/or the strong affinity of As(V) for Fe(III).

Another approach, involving the synthesis of cobalt ferrite-aggregated schwertmannite through urea hydrolysis, aimed to enhance the magnetic properties of schwertmannite for As(V) retention (Dey et al., 2014). The paramagnetic modification improved As(V) uptake with increased adsorbent mass and reaction temperature, achieving saturation in about 180 minutes.

An important in-situ remediation approach was evaluated by Chai et al. (2016) for As immobilization around a mining-affected area. This study assessed As stabilization in contaminated soil from the realgar mine site in south-central China by mixing schwertmannite with the soil at different ratios (0.5%–10%). After 7 days of contact, over 90% of As was immobilized in the soil. Water-soluble As, representing easily available As, and $NaHCO_3$-extractable As, representing bioavailable As for plant uptake, both showed significant reductions. A schwertmannite-to-soil ratio of more than 5% was sufficient to achieve As fixation, although this may vary depending on the type of soil, As leachability, and the properties of schwertmannite, which can change based on synthesis methods.

5.4 Arsenic along acid sulphate soils

In addition to As abundance near mine drainage areas, significant levels of As have also been detected in acid sulphate lowlands. Approximately 12 million hectares of acidic sulphate soils worldwide are As rich (Andriesse and van Mensvoort, 2006). This enrichment is attributed to the dissolution of As-containing pyrite (FeS_2–As), which is prevalent in these areas. Acid sulphate soils (ASSs) are formed along coastal floodplains as a result of pyrite oxidation, and since As has a strong affinity for pyrite, it frequently leads to soils rich in sulphate, iron, and arsenic.

Schwertmannite is a common ferric precipitate in such regions due to the high Fe^{3+}/Fe^{2+} ratio, SO_4^{2-} enrichment, and low pH, along other ferric oxides such as ferrihydrite, goethite, and lepidocrocite (Figure 5.2) (Burton et al., 2008; Johnston et al., 2010). Schwertmannite often undergoes reductive dissolution when anaerobic conditions develop, leading to the release of Fe(II) and the production of hydrogen sulphide (H_2S). This process results in the formation of new minerals containing iron and sulphur, while stable ferric oxides like goethite may persist. The As-rich coastal soils can leach significant amounts of As when they are re-flooded by tidal flows. This dynamic process is influenced by various factors such as microbial activity, organic matter content in aquifers, and oxygen diffusion rates. These factors govern the mineralogical redistribution of As and other elements over time and depth in these complex coastal environments. ASS can serve both as a source of As, releasing it downstream when conditions favour reductive or oxidative dissolution, and as a sink, where

FIGURE 5.2 Field photographs showing schwertmannite-rich soil and sediment formed by iron sulphide oxidation and associated acidification due to drainage of acid sulphate soil landscapes around the tidally re-flooded East Trinity wetland in northern Australia (ref. Schoepfer and Burton, 2021).

As is immobilized through adsorption onto Fe(III) oxides or iron sulphide phases, or through co-precipitation in sulphide minerals.

In ASSs, the redox distribution of As reveals a complex mixture of As species. Both As(III) and As(V), along with organo-As species, are present in the soil and pore waters. The specific distribution is influenced by redox conditions. As(V) tends to dominate under oxidizing conditions, while As(III) is prevalent in organic matter-rich reducing environments. Kinsela et al. (2011) found that in ASS of eastern Australia, organo-As species—such as methylated and thiolated As—are common in pore waters along with As(III), while As(V) remains the dominant species in the soil. Arsenic in these soils is primarily associated with residual fractions, meaning it is attached to crystalline ferric oxyhydroxides (e.g., Fe_2O_3), sulphides (e.g., FeS_2, As_2S_3), and silicates. The study also highlights that phosphate-extractable As, which represents As that can easily desorb from mineral surfaces, and ascorbate-extractable As, indicative of As bound to amorphous Fe(III) oxides like ferrihydrite and schwertmannite, only make up a minor fraction compared to the residual As. This suggests that most of the As in ASS is tightly bound to more stable mineral phases, limiting its mobility under current environmental conditions.

5.5 Antimony retention

Antimony (Sb) is recognized as a serious environmental pollutant, primarily affecting aquatic ecosystems. Its presence in the environment, especially from mining and industrial activities, poses significant risks to water quality. The main sources of Sb pollution are Sb mining and the associated processing activities, with widespread use in industries for pigments, alloys, catalysts, and glass decoration. In oxygenated environments, antimony exists predominantly in two oxidation states, i.e., Sb(V) (mainly found as $Sb(OH)_6^-$) and Sb(III) (predominantly present as $Sb(OH)_3$). Both forms are common in contaminated water sources. Regulatory bodies like the USEPA have set a strict maximum permissible limit of 0.006 mg L^{-1} for Sb in drinking water due to its toxicity (USEPA, 1984; Guo et al., 2009).

Coagulation, membrane filtration, ion exchange, and adsorption are the commonly used techniques for Sb removal where adsorption is considered particularly promising because it is cost-effective, requires minimal engineering, and is straightforward to implement. However, choosing a suitable adsorbent that provides excellent sorptive efficiency, forms strong bonds with Sb, and overcomes disposal difficulties remains a challenge. Research highlights that schwertmannite could play a vital role in mitigating Sb pollution, especially in mining vicinities where it forms naturally and can be utilized for in-situ treatment of contaminated water. Its effectiveness stems from its ability to bind strongly with Sb, particularly Sb(V), through adsorption and ion exchange processes, making it a viable option for Sb remediation.

The study by Dong et al. (2015) explored the adsorption behaviour of Sb(V) onto schwertmannite synthesized via the dialysis method. The system reached equilibrium within 5 hours of interaction between Sb(V) and schwertmannite indicating a high affinity of schwertmannite for Sb(V) at initial stages. While a maximum Sb(V) uptake of 101.7 mg g^{-1} after 24 hours was observed, its impedation with graphene has enhanced uptake efficiency significantly reaching 158.6 mg g^{-1} after 24 hours at pH 7 and 25°C. Three distinct phases of adsorption kinetics were proposed, i.e., (1) initial rapid uptake where Sb(V) rapidly binds to available external surface sites on schwertmannite, (2) intra-particle diffusion where Sb(V) diffuse into the pores of schwertmannite, and (3) the final stage slow intra-particle diffusion, indicating long-term retention of Sb(V) within the schwertmannite structure. The process was found to be endothermic, meaning Sb(V) uptake increased with rising temperature. The highest uptake was observed at lower pH levels (pH ~ 3) and it decreased as pH became alkaline (pH 9.5). This is attributed to the dominance of negatively charged $Sb(OH)_6^-$ species in alkaline conditions that are repelled by the negatively charged schwertmannite surface at pH > pH_{PZC} (pH_{PZC} ~ 5.1). Although the presence of competing anions NO_3^-, SO_4^{2-}, Cl^-, PO_4^{3-}, and SiO_4^{4-} had minimal impact on Sb(V) adsorption, HCO_3^- reduced the adsorption efficiency by more than 50% possibly due to the competition for active adsorption sites.

5.6 Chromium retention

Industrial effluents, mining operations, and improper waste disposal are the major contributors to chromium pollution. Out of the two oxidation states of chromium, i.e., Cr(VI) (which occurs either as $HCrO_4^-$ or CrO_4^{2-}, pK_a = 6.5) and Cr(III) ($Cr(OH)^{2+}$, pK_a = 4.0), Cr(VI) is more mobile and toxic that poses significant risks due to its carcinogenic nature to humans, animals, and plants compared to Cr(III). In natural environments, Cr(VI) can be reduced to Cr(III) in the presence of appropriate reductants such as organic matter, sulphides, and iron oxides. This is an important conversion because Cr(III) can precipitate out of solution, e.g., $Cr(OH)_3$, $CrFe(OH)_6$, and $CrPO_4$ (Yu et al., 2014). Chromium(VI) reduction under acidic conditions (pH < 6) is relatively faster than at neutral or alkaline pH (Lan et al., 2006). Iron oxides, including Fe(III) and Fe(II) adsorbed on soil surfaces, have been shown to promote the reduction of Cr(VI) to Cr(III) (Buerge and Hug, 1997; Lan et al., 2006).

The high reactivity and surface area of schwertmannite makes it an effective reactant for such redox reactions (Zhou et al., 2012). In the presence of 1 g L^{-1} schwertmannite, the reduction of Cr(VI) to Cr(III) by sulphide ($Na_2S{\cdot}9H_2O$) occurred 11 times faster than in its absence at pH 7.5. At higher pH values, the efficiency decreased, with reduction rates being eight times higher at pH 8 and six times higher at pH 8.8.

The effect of schwertmannite on Cr(VI) reduction in the presence of sulphides is attributed to (i) reduction of Fe(III) in schwertmannite to Fe(II) by sulphides in solution, which is the active reducing agent in this process, and (ii) reduction of Cr(VI) to Cr(III) by Fe(II) generated from schwertmannite with Fe(II) being oxidized back to Fe(III). This process continues as Fe(III) is continuously regenerated and reduced back to Fe(II) by sulphides, facilitating continuous Cr(VI) reduction until all Cr(VI) is depleted.

i. $\text{Schwertmannite} - \text{Fe}(\text{III}) + \text{sulfide} \rightarrow \text{Fe}(\text{II})$

ii. $\text{Fe}(\text{II}) + \text{Cr}(\text{VI}) \rightarrow \text{Fe}(\text{III}) + \text{Cr}(\text{III})$

In addition, the microbial reduction of Cr(VI)-loaded schwertmannite presents a promising approach for environmental remediation. Notably, only minimal release of Cr(VI) has been observed during the microbial reduction of schwertmannite (Wan et al., 2018). The reduction mechanism may involve either direct microbial reduction of Cr(VI) to Cr(III) or indirect reduction through Fe(II), which is produced via microbial reduction of Fe(III) within schwertmannite. This dual reduction pathway—mediated by either microbial activity or Fe(II)—is particularly significant, as it enables the stabilization of Cr(VI) without reintroducing it into the environment, thereby enhancing the effectiveness of detoxification strategies in contaminated systems.

The study by Jiang et al. (2012) highlights the photochemical reduction of Cr(VI) to Cr(III) using biosynthesized schwertmannite as a catalyst, enhanced by organic carboxylate anions (e.g., oxalic acid, tartaric acid). The photochemical process relies on the metal–ligand-charge-transfer (MLCT) pathway, in which Fe(III)–carboxylate complexes undergo light-driven reduction, producing reactive reductants such as Fe(II), $CO_2^{\cdot-}$, and $HO_2^{\cdot-}/O_2^{\cdot-}$. The addition of 0.6 g L^{-1} schwertmannite significantly accelerated Cr(VI) photo-reduction. Schwertmannite served as a source of Fe(III), which forms complexes with organic acids, leading to photochemical reduction reactions. Among these acids, oxalic acid has proven to be the most effective, achieving complete reduction of 100 μM Cr(VI) within 50 minutes under simulated solar light (300–800 nm). Under the same conditions, tartaric acid reduced approximately 80% of Cr(VI). The reduction efficiency followed the order: oxalic acid > acetic acid > tartaric acid. This trend is attributed to the structural differences among the carboxylate acids and their capacity to extract Fe(III) from schwertmannite. Oxalic acid, with two carboxyl groups, was the most effective in forming Fe(III)–carboxylate complexes that promoted Cr(VI) reduction. Citric acid, with three carboxyl groups and one α-hydroxy group, and tartaric acid, with two carboxyl groups and two α-hydroxy groups, were less efficient. This suggests that the number and positioning of carboxyl groups play a critical role in Fe(III) complexation and subsequent photochemical Cr(VI) reduction.

Acidic conditions ($pH < pH_{PZC}$ of 5.4) facilitate Fe(III)–oxalic acid complex formation upon adsorption of oxalic acid enhancing Cr(VI) photo-reduction. The electrode potential of the Cr(VI)/Cr(III) redox couple increases with decreasing pH, further accelerating the photo-reduction process. Increasing concentrations of carboxylate acids also favour faster Cr(VI) removal, likely due to greater availability of Fe(III)-complexing ligands.

Intra-particle diffusion plays a significant role in Cr(VI) uptake by schwertmannite, facilitating a two-step adsorption process as proposed by Zhang et al. (2019). Initially, Cr(VI) is transported onto the outer surfaces of schwertmannite, where it binds rapidly due to the easily accessible outer-sphere SO_4^{2-} groups. This is followed by diffusion into the internal sites, where Cr(VI) displaces the structurally bound SO_4^{2-}, completing the uptake process. This two-step mechanism is reflected in IR spectral data. The rapid outer-sphere adsorption is evidenced by the lowering of IR intensities associated with SO_4^{2-} vibrations at 978 cm^{-1} (outer-sphere-bound SO_4^{2-}) and 1128/611 cm^{-1} (structural SO_4^{2-}). Simultaneously, the appearance of CrO_4^{2-} vibrations at 960, 927, and 825 cm^{-1} confirms Cr(VI) incorporation into the schwertmannite structure. This process of anion exchange between schwertmannite–SO_4 and Cr(VI) has been recognized as a key mechanism for Cr(VI) retention in several studies, including Yu et al. (2014) and Fan et al. (2019). These studies observed the release of excess SO_4^{2-} from schwertmannite upon interaction with Cr(VI), reinforcing the idea that Cr(VI) displaces SO_4^{2-} during its uptake.

5.7 Retention of Cu^{2+} and Cd^{2+}

The divalent copper ion (Cu^{2+}) possesses a strong affinity for schwertmannite (Gan et al., 2015). However, the efficiency of Cu^{2+} uptake is diminished when schwertmannite is pre-loaded with trace metals like Mo(VI) or As(V). This reduction in efficiency is attributed to the preoccupancy of available surface sorption sites by these trace metals, leaving fewer sites for subsequent Cu^{2+} adsorption. Aluminium-impregnated schwertmannite has been utilized for enhanced Cu^{2+} uptake, primarily due to the increased reactivity generated by the Al^{3+} ion, which raises the density of surface functional groups (Gan et al., 2015). Aluminium itself is non-toxic to the bacterium *A. ferrooxidans* and poses minimal environmental risk, making it an effective modifier for improving metal retention capacity. Studies have shown that 90% of Cu^{2+} is removed within just 20 minutes of interaction with Al-impregnated schwertmannite, with greater uptake observed at higher Fe/Al ratios. The improved retention is attributed to both the increase in surface area and the greater density of reactive functional groups available for metal binding.

Cadmium is also strongly retained onto schwertmannite in monolayer adsorption process (Fan and Zhang, 2017). Maximum adsorption occurred at higher pH values, with near 100% uptake at pH 8 possibly owing to its negatively charged surface at higher pH values, while desorption was more efficient at lower pH levels, reaching up to 80% desorption at pH 6 owing to its positively charged surface at lower pH that reduces the electrostatic attraction to Cd^{2+}.

5.8 Mechanism of contaminant retention

5.8.1 Inner-sphere vs. outer-sphere complexation

Both Langmuir and Freundlich models have been successfully applied to describe contaminant sorption onto schwertmannite, reflecting its ability to retain ions via monolayer and multilayer adsorption mechanisms. Langmuir monolayer adsorption suggests a finite number of uniform binding sites for trace metals on the schwertmannite surface, while the Freundlich multilayer adsorption model implies that adsorption occurs on heterogeneous sites with varying energy levels (Burton et al., 2009; Antelo et al., 2012; HoungAloune et al., 2014). Sorption typically occurs through two primary mechanisms, as has been demonstrated in several studies by means of extended X-ray absorption fine structure (EXAFS) spectroscopy. As an example, $H_2AsO_4^-$ binds strongly to FeO_6 octahedra on schwertmannite surfaces in an inner-sphere mechanism (Carlson et al., 2002; Waychunas et al., 1995). This type of binding is considered a strong chemical interaction, which may eventually result in the formation of amorphous ferric arsenate when the As(V)/Fe molar ratio is high.

At low concentrations, arsenate forms bidentate binuclear inner-sphere ligands while at higher concentrations, half of the As(V) forms inner-sphere and the other half outer-sphere complexes. This process distorts the schwertmannite structure, leading to detectable shifts in pair distribution functions (d-PDFs, Figure 5.3, Carrero et al., 2017). Peaks at 1.67 and 3.28 Å

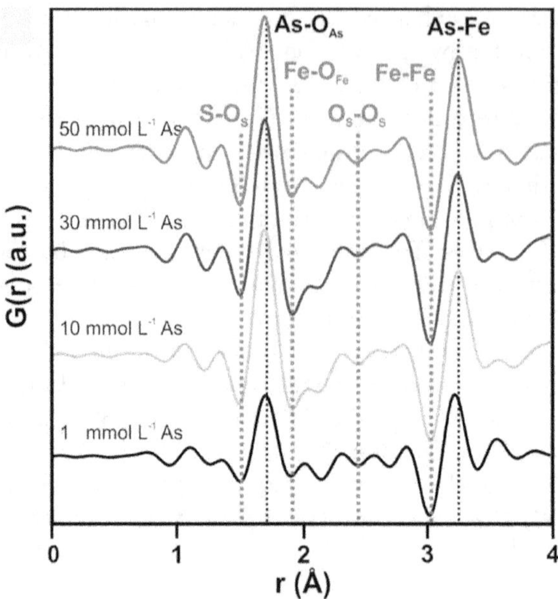

FIGURE 5.3 d-PDF of arsenate on schwertmannite in the sample loaded with 1, 10, 30, and 50 mmol L^{-1} As(V). Note the As–O and As–Fe band positions (ref. Carrero et al., 2017). Structural distortions due to As(V) sorption is reflected in Fe–O and Fe–Fe interatomic distances with negative peaks in the d-PDFs.

correspond to As–O and As–Fe distances, respectively, consistent with EXAFS data. Additionally, structural distortions in schwertmannite due to arsenate sorption are reflected by negative peaks in the d-PDFs, corresponding to Fe–O and Fe–Fe interatomic distances. Selenium(VI) sorption follows a similar pattern, with Se–O and Se–Fe distances at 1.65 and 3.34 Å, respectively (Carrero et al., 2017).

5.8.2 *The role of sulphate for the retention of contaminants*

Exchange of a contaminant for SO_4^{2-} is one of the key retention reactions. Two types of bonding are considered to be involved in these exchange reactions with inner-sphere SO_4^{2-} being more strongly bound within the structure, and outer-sphere SO_4^{2-} being more easily exchangeable. As a consequence, kinetics of exchange reactions with a series of oxoanions (SeO_3^{2-}, SeO_4^{2-}, MoO_4^{2-}, PO_4^{3-}, $Sb(OH)_6^{-}$) could be separated into an initial fast sorption process and a second slow reaction with characteristic reaction times (1/k) of ~10 minutes and ~10 hours, respectively (Marouane et al., 2021). Ligand exchange with surface-bound sulphate was proposed to be responsible for the fast reaction while intra-particle diffusion and exchange with structural sulphate for the slow process. Ion exchange of contaminants is considered to be primarily promoted by exchange with surface-bound sulphate (Fernandez-Martinez et al., 2010). Hence, schwertmannite specimens with higher SO_4^{2-} content are regarded to retain a larger amount of contaminants than schwertmannite with a lower SO_4^{2-} content, even when synthesized by the same methods (Burton et al., 2009; Paikaray et al., 2011; Antelo et al., 2012). However, the extent of ion exchange varies among different oxoanions. Arsenate exhibits relatively low retention via ion exchange, with only approximately 26% of its total uptake attributed to this mechanism.

The ion exchange coefficient for As(V)—defined as the slope of the plot of released SO_4^{2-} (mol SO_4^{2-} mol^{-1} Fe) versus adsorbed AsO_4^{3-} (mol AsO_4^{3-} mol^{-1} Fe)—was reported to be was 0.26 mol SO_4^{2-} (mol AsO_4^{3-})$^{-1}$ (Antelo et al., 2012), corresponding to an uptake of 0.196 mol As per mol Fe at pH 4.5. This low exchange coefficient suggests that As(V) retention primarily occurs through the formation of strong inner-sphere complexes with both surface-bound and structural Fe(III)–OH groups, rather than through ion exchange. Fourier transform infrared spectroscopy studies suggest that surface complexation of AsO_4^{3-} involves the replacement of OH^- groups, which causes noticeable reduction in the intensity of the OH infrared absorption band at 3358 cm^{-1} and a shift to higher wavenumbers at 3379 and 3380 cm^{-1} (Fan et al., 2019; Xiong et al., 2023). The emergence of new bands at 975 and 830 cm^{-1} confirmed the formation of an inner-sphere complex between As(V) and Fe(III)–O, which are regarded to result directly from the substitution of OH^- groups by As(V) at the schwertmannite surface (Xiong et al., 2023). In addition, part of the As uptake also appears to occur via a ligand exchange process, as indicated by the reduced intensity of the 1140 and 1110 cm^{-1} bands corresponding to SO_4^{2} vibrations.

The limited ion exchange capacity of As(V) can be attributed to its relatively large ionic radius (0.248 nm) and trivalent charge (AsO_4^{3-}), which reduce its ability to replace the smaller, divalent SO_4^{2-} anions. In contrast, the retention of chromate (CrO_4^{2-}, Cr(VI)) is predominantly governed by ion exchange, as indicated by an ion exchange coefficient of 1.23 mol CrO_4^{2-} mol^{-1} AsO_4^{3-} and a retention of 0.88 mol Cr per mol Fe at pH 4.5 (Antelo et al., 2012). This high exchange coefficient, corresponding to nearly complete exchange, suggests that CrO_4^{2-} can fully substitute for structural SO_4^{2-} in schwertmannite without disrupting its crystal structure, as demonstrated by Regenspurg and Peiffer (2005). This capacity for substitution helps explain the substantial Cr loading (up to 0.15 wt%) observed in schwertmannite precipitates downstream of the abandoned Libiola Fe–Cu sulphide mine (Marescotti et al., 2012). The effective exchange is facilitated by the comparable ionic radii and identical charges of CrO_4^{2-} and SO_4^{2-}. Consistent with these findings, Fan et al. (2019) reported similar ion exchange coefficients for Cr(VI) and As(V)—1.45 and 1.00, respectively—corresponding to uptake values of 0.54 mol Cr(VI) and 0.65 mol As(V) per mol Fe. Similar to CrO_4^{2-}, SeO_4^{2-} also exhibits +/– complete exchange with sulphate in a 1:1 exchange ratio (Carrero et al., 2017).

However, the above stoichiometric coefficients are not universal and they can vary with the SO_4^{2-} content of schwertmannite. For example, Burton et al. (2009) report a SO_4–AsO_4 exchange coefficient of 0.55 at the same pH as in the study by Antelo et al. (2012) (pH 3.0–5.5) due to excess outer-sphere-bound SO_4 that was readily exchangeable by As(V). Exchange coefficients with SO_4^{2-} can also vary significantly with pH. For example, coefficients of ~0.56–0.53 mol SO_4^{2-} per mol As(V) were determined at pH 3.0–5.5, which dropped to 0.22 at pH 9.0. A similar value of 0.62 mol SO_4^{2-} per mol As(V) was reported at pH 3.0–3.4 by Fukushi et al. (2003b). In contrast, the exchange coefficients were lower for As(III) (0.20–0.25 mol SO_4^{2-} per mol As(III) between pH 3.0 and 6.4) and remained relatively constant (0.17 and 0.16, respectively) at higher pH values (Burton et al., 2009; Liao et al. 2011). The lower exchange coefficient and the lack of pH dependence for As(III) is attributed to the larger and uncharged molecular size of the species H_3AsO_3 being the predominant As(III) species over the pH range studied, as compared to the smaller and charged $H_2AsO_4^-$ and $HAsO_4^{2-}$ anions.

5.9 The role of solution pH and surface charge in contaminant retention of schwertmannite

The highest impact on contaminant uptake is certainly exerted by the pH of the aqueous solution, which affects both the surface charge of schwertmannite and the speciation of ionic contaminants. The surface becomes negatively charged at $pH > pH_{PZC}$, which allows for repulsion of negatively charged species but attraction of positively charged species, with inverse behaviour at $pH < pH_{PZC}$. The pH values at which the pH_{PZC} is located may vary strongly among schwertmannite specimens (cf. Chapter 4) ranging from pH 3.4–3.7 (Eskandarpour et al., 2008; Hu et al., 2017), 5.4 (Liao et al., 2011), to pH 7.2 (Jönsson et al., 2005).

Uptake of anions is generally more efficient at $pH < pH_{PZC}$ where its surface becomes protonated, which attracts negatively charged oxyanion, e.g., arsenate ($HAsO_4^{2-}$) and vanadate ($H_3V_2O_7^{2-}$) (Acero et al., 2006; Fan et al., 2019). It should be noted though that formation of strong inner-sphere complexes at the schwertmannite surface or within the tunnel structure may shift the adsorption envelope of anions to even more alkaline conditions beyond the pH_{PZC} (Stumm and Morgan, 1995).

Uptake also depends on the speciation of the anions. The highest uptake of F^- by schwertmannite was determined to be around the pH_{PZC} of 3.7 or 3.4 of the schwertmannites used (Eskandarpour et al., 2008; Hu et al., 2017). At this pH, attraction of the F^- was at maximum because the surface still contained a fraction of positively charged sites. At lower pH, F^- becomes protonated that reduced the uptake and as pH increased beyond 4.0, F^- uptake decreased.

Speciation-dependent uptake was also demonstrated for As(V) (Burton et al., 2009). The uptake rate was higher at acidic pH with an uptake rate of 0.333 mol As(V) per mol Fe at pH 3 and 0.225 mol As(V) per mol Fe at pH 9. This observation was explained with the positively charged surface sites at low pH and the predominance of negatively charged As(V) species (H_3AsO_4 ($pK_a = 2.3$), $H_2AsO_4^-$ ($pK_a = 6.8$), and $HAsO_4^{2-}$ ($pK_a = 11.6$)). As the pH becomes alkaline, sorption decreases due to the negatively charged surface repelling the arsenate anions. Similar trends were observed by Song et al. (2015) who reported an As(V) loading of 183 mg g^{-1} schwertmannite at pH 3 and 143 mg g^{-1} at pH 7 within 200 minutes.

In contrast, the high pK_a values of As(III) species ($pK_{a_1} = 9.2$ and $pK_{a_2} = 12.7$) imply that sorption is higher under alkaline conditions compared to acidic environments due to the predominance of the neutral species H_3AsO_3 over a broad pH range. Hence As(III) uptake showed an opposite trend to As(V) with 0.379 mol As(III) per mol Fe at pH 9 and 0.205 mol As(III) per mol Fe at pH 3. Unlike the negatively charged As(V) species, this uncharged As(III) species does not experience repulsive forces as the pH increases favouring As (III) uptake at higher pH levels (Burton et al., 2009). Similarly, Song et al. (2015) observed a higher As(III) uptake at pH 7 (218 mg g^{-1}) than at pH 3 (46 mg g^{-1}). Liao et al. (2011) also observed an optimal As(III) uptake at pH 7.0–9.0 (Figure 5.4). At acidic pH, the low uptake of As(III) was attributed to the predominance of its neutral species in solution (H_3AsO_3). Since the schwertmannite used in their study had a pH_{PZC} of 5.4, both the schwertmannite surface and the $H_2AsO_3^-$ species were negatively charged at pH 7–9. Therefore, the higher uptake of As(III) at this pH range was explained by a specific adsorption process, presumably ligand exchange with schwertmannite-bound sulphate.

Also, the uptake of Cr(VI) by schwertmannite is strongly controlled by the aqueous pH as pH influences the speciation of Cr(VI) in solution. Below pH 6.5, Cr(VI) primarily exists as $HCrO_4^-$ and as CrO_4^{2-} at higher pH, impacting the electrostatic attraction between the negatively charged Cr(VI) species and the positively charged surface of schwertmannite and

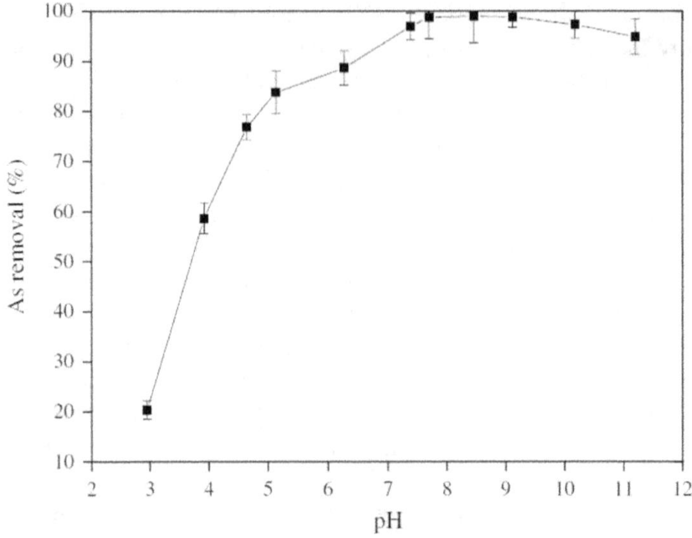

FIGURE 5.4 Effect of solution pH on As(III) uptake by biogenic schwertmannite. Initial As(III) 1.0 mg L^{-1}; equilibrium time 240 minutes (ref. Liao et al., 2011). Note the increased retention at higher pHs.

also allowing for direct exchange of CrO_4^{2-} with sulphate (Regenspurg and Peiffer, 2005; Antelo et al., 2012). Optimal uptake typically occurs at pH < pH_{PZC} (Antelo et al., 2012; Gan et al., 2015). Other metal oxoanion species such as MoO_4^{2-} or $Sb(OH)_6^-$) exhibit a similar response to pH change. Maximum retention of Mo(V) and of Cr(IV) was noted at pH 4.5 (Antelo et al., 2012), the retention decreased as the pH became more alkaline. The highest uptake rate of $Sb(OH)_6^-$ was at pH ~ 3 and the lowest one at pH 9.5 (Dong et al., 2015). In alkaline solutions, the negatively charged $Sb(OH)_6^-$ species will become repelled by the negatively charged schwertmannite surface.

The retention efficiency of positively charged metal ions, such as Cu^{2+}, Pb^{2+}, Cd^{2+}, Zn^{2+}, and Co^{2+}, shows an uptake pattern inverse to anions being strongly diminished under acidic pH conditions (Swedlund and Webster, 2001; Acero et al., 2006; Gan et al., 2015).

Also uptake rates appear to depend on pH (Cao et al., 2021). Adsorption equilibrium between schwertmannite and As(V) was reached much faster at pH 2 (~2 hours) compared to that at pH 7 (~10 hours) and pH 11 (6 hours). The rapid initial uptake of As(V) at all pH levels was attributed to electrostatic processes at the surface of the schwertmannite, which are regarded to be fast (Pierce and Moore, 1982), followed by slower interparticle diffusion into the interior of the schwertmannite particles. The kinetic model was refined by As et al. (2024) in a study on PO_4^{3-} uptake. They also observed a two-step uptake process, but the pH dependence was different with a maximum of both first-order rate constants at circum-neutral pH, which they explained in terms of competitive and pH-dependent sulphate exchange reactions.

5.10 Co-precipitation vs. adsorption/ion exchange

Unlike adsorption onto schwertmannite, retention of a contaminant during co-precipitation with Fe(III) and SO_4^{2-} will favour structural bonding and may even change the structure.

Uptake of As(V) reduced the SO_4^{2-} content in schwertmannite both by co-precipitation during synthesis of schwertmannite and by adsorption (Zhang et al., 2016). The Fe:SO_4 ratio increased in proportion with the As:Fe ratio implying replacement of SO_4^{2-} by As(V). However, the Fe:SO_4 ratio was always higher under conditions of adsorption compared to co-precipitated schwertmannite. The authors concluded that As(V) is able to replace SO_4^{2-} via ligand exchange during adsorption, while it may be structurally incorporated within schwertmannite during co-precipitation of As(V).

With increasing As content, a threshold may be exceeded during co-precipitation at which the structure of schwertmannite is collapsing. Carlson et al. (2002) observed that at an As/(As + S) molar ratio larger than 0.86, the typical schwertmannite diffraction peaks start to diminish accompanied by the simultaneous appearance of new broad peaks at 0.31 and 0.16 nm that represent amorphous ferric arsenate (FeOHAs). This aligns with findings by Zhang et al. (2016) and Wang et al. (2020) who noticed the decrease of characteristic schwertmannite peaks at 0.182 and 0.395 nm upon increasing As/Fe ratios and the appearance of new peaks at 0.263, 0.465, and 0.553 nm (Figure 5.5a). The typical IR bands for schwertmannite (e.g., at 614, 703, 1038, 1135, and 1204 cm^{-1}) decrease in intensity as As incorporation increases, while a new band at 834 cm^{-1} appears. This band is associated with As–O stretching vibrations in As–O–Fe coordination (Figure 5.5b). The increase in intensity of the 834 cm^{-1} band correlates with increased As incorporation into the structure, as reported by Zhang et al. (2016), Fan et al. (2019), and Wang et al. (2020).

5.11 External controls on contaminant retention

The mechanisms discussed in the previous subchapter refer to specific interactions between and ion and binding sites in schwertmannite. Irrespective

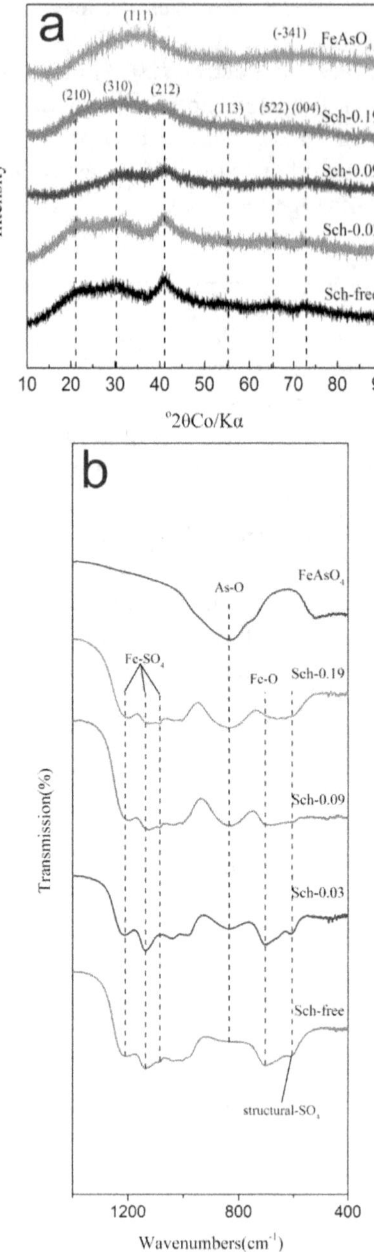

FIGURE 5.5 X-ray diffractograms (a) and infrared spectra (b) of schwertmannite specimens synthesized through rapid oxidation of Fe(II) by H_2O_2 in the absence of As(V) and at different As/Fe molar ratios. Note the changes in diffractograms like disappearance of characteristic schwertmannite peaks upon increase in As/Fe molar ratio. The synthesized amorphous $FeAsO_4$ is shown for comparison (ref. Zhang et al., 2016).

of these, external factors related to solution composition and the nature of schwertmannite exert control on the retention which we will discuss below.

5.11.1 Temperature

Trace metal uptake by schwertmannite is an endothermic process which states that the retention efficiency increases with increasing temperature. In fact, the adsorption capacity of bacterially synthesized schwertmannite for As(III) increased from ~100 mg g^{-1} at 15°C to around 120 mg g^{-1} at 35°C (Figure 5.6, Liao et al., 2011). Temperature was shown to play a significant role also in enhancing Cr(VI) uptake. At 35°C, complete removal of 100 µM Cr(VI) was achieved within 15 minutes, whereas it decreased to 87% and 89% at 15°C and 25°C at pH 8, respectively (Zhou et al., 2012). An indirect effect of temperature was observed by Houngaloune et al. (2017) where As(V) uptake increased from 17–23 mg g^{-1} for schwertmannite synthesized at 25°C to 117–128 mg g^{-1} for the one synthesized at 65°C, and which was attributed to an increase of SSA of the high-temperature sample (147.4–176.9 m^2 g^{-1}) compared the sample synthesized at 25°C (14.1–21.4 m^2 g^{-1}).

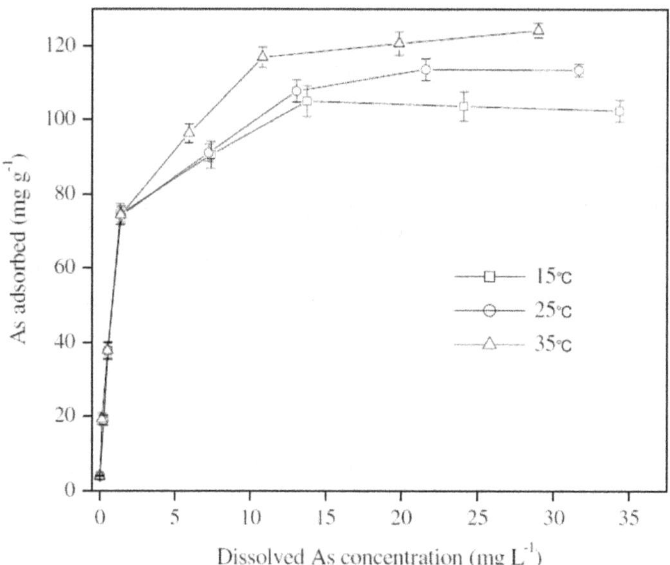

FIGURE 5.6 Uptake of As(III) by biogenic schwertmannite with respect to As concentration at 15°C, 25°C, and 35°C, respectively (ref. Liao et al., 2011). Note the increase of uptake as temperature increases.

5.11.2 Bacterial strains

Bacterial strains, particularly those belonging to A. *ferrooxidans*, exhibit variable efficiencies in oxidizing Fe(II) and trapping trace metals, especially As(III). Some strains of A. *ferrooxidans* oxidize Fe(II) rapidly, while others exhibit slower initial oxidation with sustained and more efficient Fe(II) oxidation over time (Egal et al., 2009). This difference in the capability to oxidize Fe(II) also extends to their capacity for trapping As(III) and the nature of the secondary precipitates formed during the process. In environments where A. *ferrooxidans* interacts with Fe- and SO_4-enriched mine effluents, schwertmannite is the typical reaction product forming within days. However, in media where Fe(II) and As(III) are rapidly precipitated, schwertmannite formation may be absent. When the system experiences slow Fe(II) oxidation rates and As(III) concentrations are high, the ferric arsenite–sulphate mineral tooeleite ($Fe_6(AsO_3)_4SO_4(OH)_4{\cdot}H_2O$) may form (Egal et al., 2009).

The rate of tooeleite formation varies depending on the bacterial strain, and this process is indirectly controlled by the rate of Fe(II) oxidation. Strains capable of rapid Fe(II) oxidation tend to promote faster fixation of As(III) onto the newly formed ferric precipitates. As a result, these strains are able to facilitate the quicker production of tooeleite, a process that significantly impacts As retention in such environments (Morin et al., 2003).

Research has shown that schwertmannites produced by different bacterial species can exhibit varying efficiencies in Cr(VI) retention. For instance, schwertmannite synthesized by the bacterial oxidation of Fe(II) using the strain A. *ferrooxidans* Gf showed stronger sorption of Cr(VI) compared to schwertmannite produced by the strain A. *ferrooxidans* 23270 under similar conditions (room temperature and pH 7, Zhu et al., 2013). This discrepancy in Cr(VI) retention is attributed to the capability of the A. *ferrooxidans* Gf strain to oxidize Fe(II) faster, which results in a schwertmannite with lower crystallinity more conducive to Cr(VI) sorption.

5.11.3 Specific surface area

The SSA plays a critical role on trace metal retention by schwertmannite. Dou et al. (2013) demonstrated a direct relationship between the SSA of synthesized schwertmannite specimens and their As(V) retention. Each synthesis method resulted in different morphologies (irregular, cylindrical, and spherical) with BET-SSAs of 199.4, 189.3, and 32.5 $m^2\ g^{-1}$, respectively. Uptake rates were determined at pH 7.0 with the irregularly shaped schwertmannite exhibiting the highest As(V) retention (33 $mg\ g^{-1}$), followed by the cylindrical (22 $mg\ g^{-1}$) and the spherical (4 $mg\ g^{-1}$) morphology. The study proposed a two-step sorption mechanism, i.e., external diffusion of As(V) onto the external surface of the schwertmannite particles and intra-particle diffusion where As(V) diffuse into the internal pores of schwertmannite.

5.11.4 Composition of the solution

The presence of competing anions significantly influences trace metal uptake by schwertmannite, especially when dissolved ion concentrations are relatively high (Liao et al., 2011). Monovalent anions such as Cl^- and NO_3^- have a negligible impact compared to divalent (e.g., SO_4^{2-}) and trivalent (e.g., PO_4^{3-}) anions.

In a competition experiment in the presence of AsO_4^{3-}, PO_4^{3-}, CrO_4^{2-}, and SeO_3^{2-} at equimolar concentrations at pH 7.0, Khamphila et al. (2017) found a sequence in the retention capacity with $AsO_4^{3-} > PO_4^{3-} >> CrO_4^{2-}$ and SeO_3^{2-}. Hence, the strong retention affinity of phosphate inhibits the uptake of other metal(loids) even at moderate concentrations due to competition for the same adsorption sites on schwertmannite surfaces.

The inhibitory effect of phosphate was demonstrated for the uptake rate of Cr(VI) by biosynthesized schwertmannite. The rate decreased from ~42% at pH 7 and 9 within 12 minutes, to only 2.5% in the presence of 0.1 M PO_4^{3-}. In comparison, SO_4^{2-} and NO_3^- have a relatively smaller effect (Yu et al., 2014). Also F^- showed a competitive effect, but to a lower extent. The presence of 100 µM F^- reduced Cr(VI) removal to 50% in 80 minutes, whereas complete Cr(VI) removal was achieved in the absence of F^- under the same conditions with 1 g L^{-1} schwertmannite.

Additionally, the occurrence of higher concentrations of large molecular-sized ions such as the anions of humic acids (HA) or even uncharged Si species (H_4SiO_4) inhibits retention of AsO_4^{3-} more effectively than smaller ions like Cl^-, NO_3^-, CO_3^{2-}, and SO_4^{2-} (Dou et al., 2013; Dey et al., 2014). Humic acid is particularly strongly adsorbed by schwertmannite under acidic conditions, which further hinders metal(loid) uptake (Kumpulainen et al., 2008). Xiong et al. (2023) demonstrated the order of inhibition by different ions on As(III)/As(V) uptake as follows: $HCO_3^- \approx Cl^- \approx CO_3^{2-} < SO_4^{2-} < PO_4^{3-}$. The lower effect of Cl^- and NO_3^- is due to their outer-sphere binding behaviour, meaning they do not form strong chemical bonds with the surface of schwertmannite, whereas F^-, SO_4^{2-}, and PO_4^{3-} undergo inner-sphere complexation, which involves a more direct and stronger bond to the schwertmannite surface.

5.11.5 Synthesis method

The efficiency of contaminant uptake by schwertmannite is strongly influenced by its synthesis method, even if the same method is used but with varying synthesis parameters. Differences in morphology, SSA, and the amount of exchangeable SO_4^{2-}/OH^- on schwertmannite surfaces and in its tunnel structure affect its adsorption capacity. A study by Cao et al. (2021) highlights such variability in As(V) uptake. The researchers compared the oxidation of $FeSO_4$ by H_2O_2 with oxidation by $KMnO_4$ at varying oxidant dosages. As(V) adsorption decreased as the oxidant dosage increased, likely due to the corresponding growth in schwertmannite particle size. This effect was more pronounced in schwertmannite synthesized using $KMnO_4$. Similarly, Xiong et al. (2023) observed that biosynthesized schwertmannite exhibited a higher uptake of As(III) and As(V) compared to the schwertmannite synthesized by chemical oxidation. The difference in uptake rate increased when both the adsorbate concentration and the amount of adsorbent increased.

As(V) uptake was relatively low (~22 mg g^{-1}) at pH 2, irrespective of the synthesis method, while it increased substantially at pH 7 with clear differences between the specimens. Schwertmannite synthesized in the presence of $KMnO_4$ had a much higher uptake rate (~80 mg g^{-1}) compared to schwertmannite synthesized in the presence of H_2O_2 (~40 mg g^{-1}). The As(V) uptake rates and the SO_4^{2-}–As(V) exchange coefficients depended on the amount of the oxidant. At pH 11, the uptake rate was ~40 mg g^{-1} for both types.

Also, the rate of addition of an oxidant appears to have an influence on uptake rates. At a slow H_2O_2 addition rate, Cr(VI) uptake was promoted, while its retention was reduced at a faster addition rate (Zhang et al, 2019). This observation was attributed to changes in the SSA and the pore volume of schwertmannite. Slow addition increased both SSA and pore volume and facilitated adsorption through an ion exchange mechanism involving the structural SO_4^{2-} groups of schwertmannite.

5.12 Schwertmannite regeneration and modifications

Application of schwertmannite as a sorbent material for remediation purposes has driven research to regenerate schwertmannite after its application and to evaluate its efficiency towards contaminant retention with the objective to reduce the cost of wastewater treatment.

Schwertmannite demonstrates significant potential for As(V) retention even after multiple regeneration cycles with NaOH. Granular schwertmannites, in both irregular and cylindrical forms, maintained an uptake efficiency of over 91% of the original As(V) adsorption capacity to remove a solution concentration of 10 mg L^{-1} As(V) at pH 7 after four regeneration cycles without detectable changes in structure (Dou et al., 2013). Regeneration with 0.1 M NaOH was more effective than that with 1 M

NaOH possibly because of destruction of the active sites or structural disintegration of the schwertmannite at higher NaOH concentrations. Similar results were obtained for schwertmannite amended with cobalt ferrite with an As(V) retention efficiency of 98% after regeneration (Dey et al., 2014). Similar efficiencies for the treatment of As(III)-containing solutions were reported after regeneration of biosynthesized schwertmannite with NaOH (bacterial oxidation of $FeSO_4 \cdot H_2O$ with *A. ferrooxidans*, Yue and Li-Xiang, 2013).

Regeneration of Cr(VI)-loaded schwertmannite using H_2SO_4 at pH 2 has yielded a 99% uptake efficiency (Zhang et al., 2019). The uptake efficiency decreased slightly over time, yet four cycles of regeneration still yielded over 91% Cr(VI) removal from solution containing 2 mM of Cr(VI).

Regeneration of schwertmannite synthesized following the $KMnO_4$ and MnO_2 oxidation pathways (cf. Chapter 3) to remove F^- was successfully applied (Zhu et al., 2020). The schwertmannite was synthesized in the presence of F^- so that this ion was initially co-precipitated. The regeneration process involved washing of a F^--containing schwertmannite with dilute H_2SO_4 at pH 1.5 to remove adsorbed fluoride. After cleaning, the schwertmannite was reused for adsorptive removal to treat a solution containing 15 mM F^- at pH 4, maintaining a removal efficiency of ~83% over five regeneration cycles, without significantly altering the schwertmannite's structure.

Schwertmannite has demonstrated a strong capacity for Cu^{2+} uptake, even after its regeneration (Gan et al., 2015). The regeneration process involves removal of previously adsorbed Cu^{2+} ions using acidified water at pH 2.0 and its subsequent reuse in adsorbing the same metal.

Modification of schwertmannite with specific reactants turned out to improve the removal efficiency. Co-precipitation of Cu^{2+} with schwertmannite during dialysis-controlled synthesis has enhanced As(V) uptake by nearly 30% compared to a untreated control (Antelo et al., 2013). The increase in As(V) retention is assumed to be related to the partial substitution of Cu^{2+} for Fe^{3+} within the schwertmannite structure and the presence of Cu^{2+} enhanced As(V) retention. The variation of the solid-phase Cu^{2+} content did not significantly impact the amount of As(V) uptake. As the schwertmannite aged, its retention capacity significantly declined, likely due to the progressive transformation into goethite. A similar trend was observed by HoungAloune et al. (2014) in the presence of Cu^{2+} in solution. Conversely, Fe^{2+} substantially lowered As(V) uptake, for instance, at an Fe^{2+} concentration of 1 mM, As(V) uptake was 149 mg g^{-1}. Uptake dropped to 58 mg g^{-1} when Fe^{2+} concentration was increased to 100 mM. This decline can be attributed to the catalytic effect of $Fe(II)_{aq}$ in promoting transformation of schwertmannite to goethite. Notably, the simultaneous presence of both Fe^{2+} and Cu^{2+} enhanced As(V) uptake, indicating a synergistic effect between the two metals.

5.13 Application of schwertmannite to remove non-mining contaminants

5.13.1 Phosphate retention

Phosphate is an oxoanion that is typically not occurring in AMD environments. Several studies have been interested in the use of iron-containing sludge products for phosphorus removal from water (Wei et al., 2008; Münch et al., 2024). The potential for removal of phosphate with schwertmannite as residual materials is currently under debate (e.g., As et al., 2024).

Schwertmannite has a high capacity for PO_4^{3-} retention, especially under acidic pH conditions. This capacity is attributed to its pH_{PZC}. At pH < pH_{PZC}, the positively charged schwertmannite surface facilitates electrostatic attraction of negatively charged phosphate species (PO_4^{3-}, HPO_4^{2-}, $H_2PO_4^-$). On the other hand, the repulsive forces at pH > pH_{PZC} between PO_4^{3-} ions and the schwertmannite and the increasing OH^- density competes with PO_4^{3-} for adsorption sites and reduces its uptake efficiency.

In a comparative study of 63 adsorbents, Ooi et al. (2017) found that schwertmannite exhibited the strongest affinity towards PO_4^{3-} and tuned out to be most effective in treating a dilute bicarbonate buffered solution (3 mg P L^{-1}), regardless of its synthesis method. An increase in schwertmannite mass also leads to higher PO_4^{3-} uptake, as more adsorption sites become available. A study by Khamphila et al. (2017) reported that PO_4^{3-} retention (0.934 mmol g^{-1}) was higher than that of Cr(VI) (0.723 mmol g^{-1}) and of Se(IV) (0.313 mmol g^{-1}). Field-scale observations by Boukemara et al. (2017) support these findings, showing that schwertmannite-dominated mine drainage sediments retain PO_4^{3-} better than sediments rich in jarosite and quartz.

Two primary mechanisms govern the PO_4^{3-} uptake behaviour of schwertmannite, i.e., (1) electrostatic attraction between the positively charged schwertmannite surface and PO_4^{3-} ions at pH < pH_{PZC} and electrostatic repulsion at pH > pH_{PZC} that hinders uptake; (2) ligand exchange between PO_4^{3-} and schwertmannite–SO_4^{2-}, which results in a strong binding of PO_4^{3-}, making it a part of the schwertmannite structure. The ligand exchange mechanism is significant, especially when electrostatic repulsion is present (at pH > pH_{PZC}) and is evidenced by the reduction in the SO_4 IR-vibration band intensities (1103, 1072, and 979 cm^{-1}) after PO_4^{3-} adsorption (Ooi et al., 2017).

Phosphate sorption to schwertmannite, however, is complex and likely involves ligand exchange for inner- and outer-spherically coordinated sulphate groups, both on the surface and in the tunnel structure of the mineral. A strong correlation between PO_4^{3-} sorption and SO_4^{2-} release was found by As et al. (2024). The kinetics of PO_4^{3-} sorption and SO_4^{2-} release were faster at higher pH levels and maximum PO_4^{3-} sorption was found at pH 6 (1.7 mmol PO_4^{3-} g^{-1}), which decreased to 1.5 and 1.2 mmol PO_4^{3-} g^{-1} at pH 3 and 8, respectively.

^{57}Fe-Mössbauer analyses of schwertmannite demonstrated an initial transformation of schwertmannite at neutral to alkaline pH, characterized by a rise in a partially ordered sextet area. This change was interpreted as an increase in crystallinity resulting from the transition from Fe–SO_4 to Fe–O domains. The emerging phase differed from the original schwertmannite, but did not represent a complete change to a new crystalline phase, thus indicating a proto-transformation of schwertmannite. This pH-induced proto-transformation was inhibited in the presence of phosphate. We concluded that the phosphate sorption rate and maximum as well as the proto-transformation of schwertmannite were strongly affected by the mineral's affinity for sulphate. Sorption to schwertmannite should primarily be regarded as a competitive exchange reaction between the sorbing oxyanion and the bound sulphate. As a result, the highest phosphate sorption occurred at circum-neutral pH, in stark contrast to non-sulphate-containing Fe(III) (oxyhydr)oxides, where phosphate sorption is highest at acidic pH.

Schwertmannite's magnetic properties also play a critical role in PO_4^{3-} removal. As a paramagnetic material, its magnetic susceptibility can be several orders of magnitude higher than that of ordinary paramagnetic substances (ranging from 10^{-3} to 10^{-8}). In a study by Eskandarpour et al. (2006), schwertmannite synthesized through urea hydrolysis exhibited a magnetic susceptibility of 4×10^{-4} with a pH_{PZC} of 4. This magnetic behaviour significantly influences PO_4^{3-} retention, as magnetic filtration techniques enhance the removal efficiency. The process allows for better separation of schwertmannite particles and boosts PO_4^{3-} uptake without requiring magnetic seeding.

5.13.2 Fluoride retention

Fluoride (F^-) pollution is a major environmental concern, particularly in wastewater originating from industries such as metallurgy, soldering, coal burning, and other chemical processes. These wastewaters often have a pH below 5 and contain high concentrations of fluoride ions (F^-). Fluorite (CaF_2) and fluoroapatite ($Ca_5(PO_4)_3F$) are the most common naturally occurring fluoride-bearing minerals. These minerals can release F^- into groundwater through geochemical processes (Eskandarpour et al., 2008).

Common geochemical reactions responsible for the release of F^- into groundwater include (Eqs. 5.1 and 5.2):

$$CaF_2 + Na_2CO_3 \rightarrow CaCO_3 + 2F^- + 2Na^+ \quad \text{(Eq. 5.1)}$$

$$CaF_2 + 2NaHCO_3 \rightarrow CaCO_3 + 2F^- + 2Na^+ + H_2O + CO_2 \quad \text{(Eq. 5.2)}$$

Adsorptive removal of F^- has special importance due to its economic feasibility and non-engineered designs by use of activated alumina, granular ceramic, surface-tailored zeolites, granular red mud, and carbon nanotubes (Mohapatra et al., 2009; Bhatnagar et al., 2011). Schwertmannite is an excellent sorbent for fluoride treatment, outperforming many of the aforementioned adsorbents. Studies by Goswami and Purkait (2014) demonstrated that schwertmannite into which magnetite nanoparticles were embedded rapidly absorbed F^-, reaching equilibrium within just 10 minutes under normal atmospheric conditions. Longer contact times between schwertmannite and F^--rich aqueous media, as well as higher amounts of schwertmannite, result in higher F^- uptake. Fluoride retention was higher if F^- was co-precipitated during schwertmannite synthesis method compared to adsorption only, due to the enhanced availability of adsorption sites.

Fluoride removal by schwertmannite synthesized by oxidation of $FeSO_4$ with $KMnO_4$ or MnO_2 has been shown to achieve rapid equilibrium within 30 minutes of contact at pH 4 and 20°C, with maximum removal efficiencies of 93% and 80% of a 10 mM F^- solution, respectively.

The decrease in the IR band intensities for OH^- at 3430 cm^{-1} and for SO_4^{2-} at 1180, 970, and 618 cm^{-1} suggests that F^- can replace OH^- and SO_4^{2-} ions bound to Fe sites both internal and at the surface, as shown in Eq. 5.3 (Hu et al., 2017; Zhu et al., 2020).

$$\equiv FeOH^{-1/2} + H^+ + F^- \leftrightarrow \equiv FeF^{-1/2} + H_2O \quad \text{(Eq. 5.3)}$$

with ≡FeOH and ≡FeF denoting schwertmannite Fe binding sites.

Further insights into the adsorption process are provided by X-ray photoelectron spectroscopy (XPS) data (Figure 5.7, Zhu et al., 2020). The Fe 2p XPS peaks of schwertmannite at 713.7 eV (Fe(III)–SO_4), 711.8 eV (Fe(III)–OH), and 710.9 eV (Fe(III)–O) decreased in intensity after F^- incorporation. A new peak at 714.8 eV appeared, indicating the formation of Fe(III)–F complexes. The reduction of the 711.8 eV peak, corresponding to Fe(III)–OH, was more significant than that of the 713.7 eV peak (Fe(III)–SO_4), suggesting that F^-–OH^- exchange plays a larger role than SO_4^{2-}–F^- exchange in F^- retention. This observation is confirmed by mass balance calculations, which show that ~0.75 mmol g^{-1} of SO_4^{2-} and ~1.65 mmol g^{-1} of OH^- were displaced by F^- during adsorption.

Schwertmannite synthesized through the oxidation of $FeSO_4$ with $KMnO_4$ retained a higher amount of F^- compared to that synthesized by use of MnO_2 (Zhu et al., 2020). The enhanced fluoride removal in the $KMnO_4$ oxidation pathway is attributed to the higher content of OH^- and SO_4^{2-} groups in this schwertmannite specimen, which facilitates F^- retention. Among these, surface-bound SO_4^{2-} plays a minor role in F^- removal compared to OH^- that can be easily exchanged for F^-. Tunnel SO_4^{2-} contributes even less to the exchange process. Thus, the higher concentration of OH^- in $KMnO_4$-synthesized schwertmannite enhances its effectiveness for F^- removal.

The ability of F^- to replace OH^- ions in the schwertmannite structure during adsorption can be explained by their similar charge and comparable

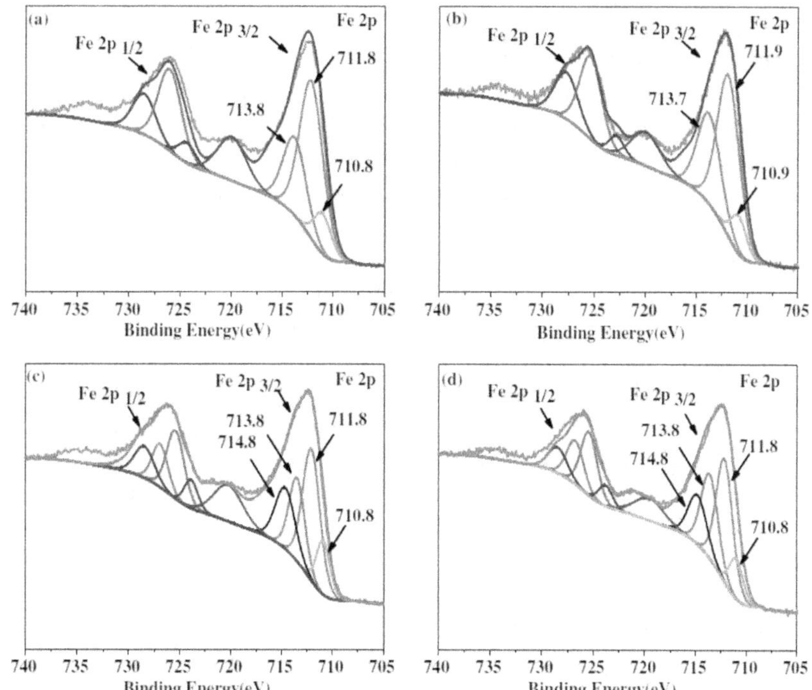

FIGURE 5.7 XPS signals at the Fe 2p region for pure schwertmannites synthesized through oxidation of $FeSO_4$ by $KMnO_4$ (a) and MnO_2 (b) and their respective Fe 2p spectra after F^- incorporation (c and d, respectively) at pH 4 (ref. Zhu et al., 2020). Note the decrease in several peak intensities after F^- incorporation and the appearance of a peak at 714.8 eV that was assigned to a Fe(III)–F complex.

ionic radii—F^- (0.119 nm) closely matches OH^- (0.110 nm). This replacement mechanism enables efficient F^- uptake under favourable conditions. However, at alkaline pH levels, beyond the pH_{PZC} of schwertmannite, OH^- becomes increasingly competitive for adsorption sites, leading to the desorption of F^-.

Desorption studies demonstrate that 85% of F^- was released from schwertmannite when the pH was raised to 11.5, indicating that F^- binding within schwertmannite weakens significantly under alkaline conditions due to the growing competition of OH^- ions at high pH (Goswami and Purkait, 2014). A comparable desorption rate of 94% was observed under acidic conditions (pH 2–3, Eskandarpour et al., 2008; Goswami and Purkait, 2014).

5.14 The role of schwertmannite in the degradation of organic pollutants

Organic pollutants, especially from industrial activities like textile manufacturing, can contribute large amounts of dyes such as acid orange 7 (AO7) into surrounding water bodies. Photocatalytic degradation, particularly involving iron-based systems, is an efficient technique to address these pollutants, as it involves generating reactive species like hydroxyl radicals ($^{\cdot}OH$) that can rapidly oxidize organic contaminants (Guo et al., 2015). Studies on Fe(III)–organic acid systems have shown that $^{\cdot}OH$ radicals form via a ligand-to-metal charge transfer mechanism, which is central to the photodegradation process. In the presence of oxygen (O_2), H_2O_2 is generated, which further reacts with Fe(II) to regenerate Fe(III) and produce more hydroxyl radicals through the Fenton reaction. These $^{\cdot}OH$ radicals are highly effective in oxidizing organic pollutants (Guo et al., 2015).

Schwertmannite has shown promising effects in photocatalytic degradation due to its Fe(III) content. In the presence of organic acids, such as citric acid, the formation of Fe(III)–organic acid complexes enhances the photodegradation efficiency. Specifically, natural organic matter (NOM) components, such as humic acids (HAs) and fulvic acids (FAs), which are adsorbing onto iron surfaces via their carboxyl and hydroxyl groups. Adsorption is more prominent at low pH due to ligand exchange and the formation of surface complexes, where carboxyl groups are more active at low pH, while hydroxylic groups become important at high pH.

Research by Guo et al. (2015) demonstrated that schwertmannite significantly enhances the photodegradation of AO7 in the presence of citric acid. At pH 5, schwertmannite alone achieved 28.4% removal of AO7, whereas citric acid alone only removed 5.8%. Citric acid plays a crucial role by forming an Fe(III)–citrate complex, which possesses photocatalytic activity under UV irradiation. This complex enhances $^{\cdot}OH$ production through the photo-Fenton reaction, where surface-adsorbed ≡Fe(III)–citric acid and dissolved Fe(III)–citric acid complexes lead to the generation of $CO_2^{\cdot-}$ and $O_2^{\cdot-}$ radicals. These radicals further facilitate the production of H_2O_2 and $^{\cdot}OH$ radicals, accelerating the degradation of AO7.

While Fe(III)–citrate (or alternatively, Fe(III)–oxalate, cf. below) are effective for photodegradation, the combination of schwertmannite and citric acid is particularly efficient due to the rapid generation of $^{\cdot}OH$ and the strong interactions between Fe(III) and organic acids. When combined, the system achieved >97% removal within 2 hours, illustrating the synergistic effect (Guo et al., 2015). The enhanced photodegradation occurs via the dissolution of schwertmannite, producing various iron complexes (Fe(III), $Fe(OH)^{2+}$, and $Fe_2(OH)_2^{4+}$), which promote the generation of $^{\cdot}OH$ radicals. The most significant contributor to $^{\cdot}OH$ production is the $Fe(OH)^{2+}$ complex (Guo et al., 2015). This synergy makes schwertmannite a valuable material for photocatalytic remediation of organic pollutants in industrial wastewater. Removal of A07 was enhanced at higher schwertmannite mass and at an aqueous $pH < pH_{PZC}$. An acidic pH favours adsorption of the citrate anion and the formation of the Fe(III)–citrate complex and this is essential for $^{\cdot}OH$ formation.

The potential of schwertmannite to catalyze photochemical degradation has also been demonstrated for methyl orange (MO) in the presence of oxalate (Wu et al., 2012). While schwertmannite or oxalate alone did not significantly degrade MO. They showed a high photochemical reactivity when combined, which allowed for the formation of reactive Fe(III)–oxalate complexes. Similar to citrate, the reaction is assumed to be triggered by adsorption of oxalate onto the surface of schwertmannite, creating ≡Fe(III)–oxalate complexes that lead to ligand–metal charge transfer. This process produces Fe(II) and $CO_2^{\cdot-}$ radicals, which react with oxygen to generate $O_2^{\cdot-}$ radicals, further producing H_2O_2 and $^{\cdot}OH$ radicals.

The photodegradation process is efficient and peaks within 5 minutes for Fe(II) production and 10 minutes for hydroxyl radical generation, with MO being ultimately oxidized by these radicals. Excess oxalate can consume more $^{\cdot}OH$ radicals, reducing the degradation efficiency, and higher MO concentrations extend the reaction time but increase the total amount of MO removed.

Schwertmannite also acts as an effective catalyst for the degradation of organic pollutants like phenol in the presence of H_2O_2 through a dark heterogeneous Fenton-like reaction (Wang et al. 2013; Yan et al., 2017), i.e., in the absence of light. In a dark heterogeneous Fenton-like reaction, H_2O_2 is reducing solid-phase Fe^{3+} to Fe^{2+}. The rate of Fe^{2+} formation is initially low. Fe^{2+} further reacts with H_2O_2 upon formation of OH radicals and Fe^{3+}.

The reaction is strongly dependent on pH and most preferential at low pH. Among other Fe(III) sources also schwertmannite provided promising results.

Degradation of phenol followed a two-step mechanism (Yan et al., 2017), i.e., (1) the initial induction phase where the degradation rate is slow but accelerates with increased H_2O_2 concentration; (2) rapid phenol degradation occurs thereafter with optimal results achieved at pH 3.0–4.5 at which the maximum production of OH radicals occurred and at which phenol degradation was most effective, while higher pH levels reduce their efficiency. Additionally, increasing the schwertmannite load and temperature boosts phenol degradation by providing more surface area for the reaction (Figure 5.8).

A considerable fraction of phenol degradation appears to be caused by the formation of $^{\cdot}OH$ radicals on schwertmannite surface (Wang et al., 2013) that provide a reaction scheme according to which phenol is eventually degraded to intermediates like quinones, catechol, oxalic acid, and acetic acid, which themselves may stimulate reduction of Fe(III) to Fe(II) thereby becoming oxidized and transformed to H_2O and CO_2 and maintaining the catalytic activity of the Fe(II)/Fe(III) cycling process (Eqs. 5.4–5.11).

Solution based $^{\cdot}OH$ radical formation

$$\text{Schwertmannite} \xrightarrow{\text{dissolution}} Fe^{3+} + SO_4^{2-} \qquad \text{(Eq. 5.4)}$$

$$Fe(III) + H_2O_2 \rightarrow Fe(II) + H^+ + {}^{\cdot}HO_2 \qquad \text{(Eq. 5.5)}$$

$$Fe(II) + H_2O_2 \rightarrow Fe(III) + OH^{\cdot -} + OH \qquad \text{(Eq. 5.6)}$$

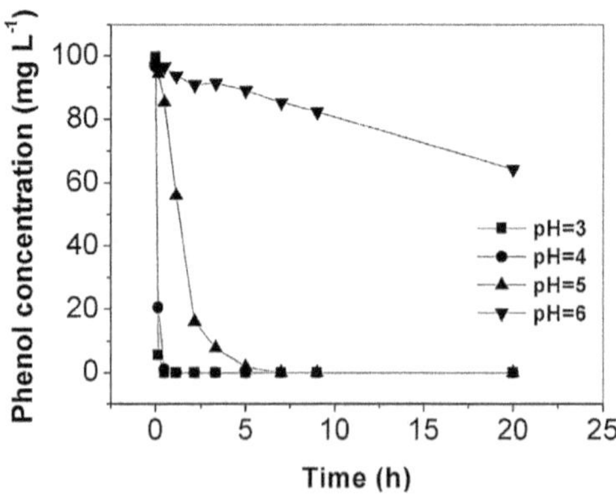

FIGURE 5.8 Removal of phenol in a heterogeneous Fenton-like reaction with schwertmannite and H_2O_2 at different pH conditions (ref. Wang et al., 2013). Note the strong pH dependency of this reaction.

Surface based $^{\cdot}OH$ radical formation

$$\equiv Fe(III) + H_2O_2 \rightarrow \equiv Fe(III) - H_2O_2 \quad \text{(Eq. 5.7)}$$

$$\equiv Fe(III) - H_2O_2 \rightarrow \equiv Fe(II) + ^{\cdot}HO_2 + H^+ \quad \text{(Eq. 5.8)}$$

$$\equiv Fe(III) + ^{\cdot}HO_2 \rightarrow \equiv Fe(II) + O_2 + H^+ \quad \text{(Eq. 5.9)}$$

$$\equiv Fe(II) + H_2O_2 \rightarrow \equiv Fe(III) + ^{\cdot}OH + OH^- \quad \text{(Eq. 5.10)}$$

$$\text{Phenol} + ^{\cdot}OH \rightarrow \frac{\text{quinones}}{\text{cathechol}} \rightarrow \frac{\text{oxalic}}{\text{fulfic}} \text{acid} + H_2O + CO_2 \quad \text{(Eq. 5.11)}$$

The presence of Cl^-, NO_3^-, and SO_4^{2-} ions had little impact on phenol degradation, although H_2O_2 oxidation was slightly hindered (Wang et al., 2013). More than 98% phenol degradation was still achieved even after 12 regeneration cycles, confirming that schwertmannite maintained its catalytic potential. Schwertmannite's reusability was further demonstrated, with no significant loss of efficiency.

In a study on heterogeneous Fenton degradation, more than 92% degradation was achieved in the presence of schwertmannite and H_2O_2 at pH 3 (Duan et al., 2016). Increasing the schwertmannite content enhanced nitrobenzene removal, while increasing nitrobenzene concentration slowed the degradation rate. The concentration of H_2O_2 also influenced the process, where application of H_2O_2 in excess (>2000 mg L^{-1}) reduced degradation efficiency due to the formation of $HO_2^{\cdot}$ radicals, which are less-effective oxidants compared to $^{\cdot}OH$ radicals.

Notably, the formation of intermediate compounds, such as three nitrophenol isomers, suggests that nitrobenzene does not break down directly into CO_2 and H_2O. Instead, intermediates form during the degradation process and are subsequently oxidized. This is consistent with findings from Wang et al. (2013), where intermediate products were formed before the complete degradation of phenol. Reusability tests also showed that schwertmannite remained highly efficient (>96%) for at least five cycles with simple washing using DI water (Duan et al., 2016).

5.15 Summary

Schwertmannite is a highly effective adsorbent for various pollutants commonly found in mine drainage and ASS environments, including arsenic (As), chromium (Cr), lead (Pb), zinc (Zn), copper (Cu), and others. It is a dominant secondary precipitate under acidic pH conditions, characterized by a high surface area and poorly crystalline (amorphous) structure.

Schwertmannite often co-occurs with other ferric oxyhydroxides such as ferrihydrite, jarosite, and goethite.

In this chapter, we summarize the natural attenuation processes affecting the most frequently occurring contaminants in AMD-impacted environments, with particular emphasis on arsenic (As), antimony (Sb), chromium (Cr), as well as the cations copper (Cu^{2+}) and cadmium (Cd^{2+}). Contaminant retention mechanisms involve both inner-sphere and outer-sphere complexation, occurring within the tunnel-like structure of schwertmannite as well as on its surface. A critical factor in these mechanisms is the presence of structurally bound sulphate (SO_4^{2-}), which participates in ligand exchange reactions. This process is the dominant retention pathway for oxyanions of similar size and charge, such as chromate (CrO_4^{2-}) and selenate (SeO_4^{2-}), but also plays an important role in the uptake of arsenate (AsO_4^{3-}).

Among the environmental variables, pH exerts the strongest influence on contaminant uptake. It affects both the surface charge of schwertmannite and the chemical speciation of dissolved contaminants. When the pH exceeds the pH_{PZC}, the schwertmannite surface becomes negatively charged, favouring the adsorption of cationic species and repelling anionic ones; the opposite occurs at pH values below pH_{PZC}. However, the exact pH_{PZC} value can vary substantially among different schwertmannite samples.

Retention mechanisms may differ significantly during co-precipitation processes, in which contaminants are incorporated during the formation of schwertmannite via Fe(III) and SO_4^{2-}. In such cases, structural incorporation is favoured and may even alter the crystal structure of schwertmannite.

The formation and retention behaviour of schwertmannite are strongly influenced by environmental factors, particularly temperature and solution composition. Major formation pathways include the oxidation of Fe(II) in sulphate-rich solutions and the dilution of strongly acidic, Fe(III)-containing waters, followed by hydrolysis in the presence of sulphate. Additionally, microbial activity can play a crucial role in these processes and is considered an important factor in schwertmannite formation and contaminant retention.

Due to its high sorption capacity, schwertmannite has been applied as a sorbent material in environmental remediation. The efficiency of schwertmannite after application can be influenced by its regeneration and modification. Furthermore, there is growing interest in its ability to retain pollutants not typically associated with AMD, such as fluoride (F^-) and phosphate (PO_4^{3-}). These species are retained through electrostatic attraction under $pH < pH_{pzc}$ conditions and via ligand exchange with SO_4^{2-}.

In addition to its use as a sorbent, schwertmannite has also been employed as a source of Fe(III) for both photochemical and heterogeneous dark Fenton-like degradation of organic pollutants, including dyes and phenol. In these applications, schwertmannite has demonstrated excellent catalytic performance.

References

Acero P, Ayora C, Torrentó C and Nieto J (2006). The behavior of trace elements during schwertmannite precipitation and subsequent transformation into goethite and jarosite. *Geochimica et Cosmochimica Acta* 70: 4130–4139.

Alarcón R, Gaviria J and Dold B (2014). Liberation of adsorbed and coprecipitated arsenic from jarosite, schwertmannite, ferrihydrite, and goethite in seawater. *Minerals* 4: 603–620.

Andriesse W and van Mensvoort MEF (2006). Acid sulfate soils: Distribution and extent. In: Rattan L (Ed), *Encyclopedia of soil science*, vol. 1. 2nd Ed. CRC Press, pp. 14–19.

Antelo J, Fiol S, Gondar D, López R and Arce F (2012). Comparison of arsenate, chromate and molybdate binding on schwertmannite: Surface adsorption vs. anion-exchange. *Journal of Colloid and Interface Science* 386: 338–343.

Antelo J, Fiol S, Gondar D, Pérez C, López R and Arce F (2013). Cu(II) incorporation to schwertmannite: Effect on stability and reactivity under AMD conditions. *Geochimica et Cosmochimica Acta* 119: 149–163.

As K, Peiffer S, Uhuegbue PO, Joshi P, Kappler A, Marouane B and Hockmann K (2024). Sulfate affinity controls phosphate sorption and the prototransformation of schwertmannite. *Chemical Geology* 653: 122043.

Baleeiro A, Fiol S, Otero-Fariña A and Antelo J (2018). Surface chemistry of iron oxides formed by neutralization of acidic mine waters: Removal of trace metals. *Applied Geochemistry* 89: 129–137.

Bhatnagar A, Kumar E and Sillanpää M (2011). Fluoride removal from water by adsorption—A review. *Chemical Engineering Journal* 171: 811–840.

Bigham JM, Schwertmann U and Pfab G (1996a). Influence of pH on mineral speciation in a bioreactor stimulating acid mine drainage. *Applied Geochemistry* 11: 845–849.

Bigham JM, Schwertmann U, Traina SJ, Winland RL, Wolf M (1996b). Schwertmannite and the chemical modeling of iron in acid sulfate waters. *Geochimica et Cosmochimica Acta* 60: 2111–2121.

Boukemara L, Boukhalfa C, Azzouz S, Reinert L, Duclaux L, Amrane A and Szymczyk A (2017). Characterization of phosphorus interaction with sediments affected by acid mine drainage-relation with the sediment composition. *International Journal of Sediment Research* 32: 481–486.

Buerge IJ and Hug SJ (1997). Kinetics and pH dependence of chromium(VI) reduction by iron(II). *Environmental Science & Technology* 31: 1426–1432.

Burgos WD, Borch T, Troyer LD, Luan F, Larson LN, Brown JF, Lambson J and Shimizu M (2012). Schwertmannite and Fe oxides formed by biological low-pH Fe(II) oxidation versus abiotic neutralization: Impact on trace metal sequestration. *Geochimica et Cosmochimica Acta* 76: 29–44.

Burton ED, Bush R, Sullivan LA, Johnston SG and Hocking RK (2008). Mobility of arsenic and selected metals during re-flooding of iron- and organic-rich acid-sulfate soil. *Chemical Geology* 253: 64–73.

Burton ED, Bush RT, Johnston SG, Watling K, Hocking RK, Sullivan LA and Parker GK (2009). Sorption of arsenic(V) and arsenic(III) to schwertmannite. *Environmental Science & Technology* 43: 9202–9207.

Cao Q, Chen C, Li K, Sun T, Shen Z and Jia J (2021). Arsenic(V) removal behavior of schwertmannite synthesized by $KMnO_4$ rapid oxidation with high adsorption capacity and Fe utilization. *Chemosphere* 264: 128398.

Carlson C, Bigham JM, Schwertmann U, Kyek A and Wagner F (2002). Scavenging of As from acid mine drainage by schwertmannite and ferrihydrite: A comparison with synthetic analogues. *Environmental Science & Technology* 36: 1712–1719.

Carrero S, Fernandez-Martinez A, Pérez-López R, Poulain A, Salas-Colera E and Nieto JM (2017). Arsenate and selenate scavenging by basaluminite: Insights into the reactivity of aluminum phases in acid mine drainage. *Environmental Science & Technology* 51: 28–37.

Chai L, Tang J, Liao Y, Yang Z, Liang L, Li Q, Wang H and Yang W (2016). Biosynthesis of schwertmannite by *Acidithiobacillus ferrooxidans* and its application in arsenic immobilization in the contaminated soil. *Journal of Soils and Sediments* 16: 2430–2438.

Chapman BM, Jones DR and Jung RF (1983). Processes controlling metal ion attenuation in acid mine drainage streams. *Geochimica ef Cosmwhimica Acta* 47: 1957–1973.

Cullen WR and Reimer KJ (1989). Arsenic speciation in the environment. *Chemical Review* 89: 713–764.

Dakos Z, Kupka D, Kovařik M, Jablonovská K, Krištúfek V and Achimovičová M (2012). Secondary iron minerals present in amd sediments from smolník abandoned mine. *Nova Biotechnologica et Chimica* 11–2*:* 87–92.

Dey A, Singh R and Purkait MK (2014). Cobalt ferrite nanoparticles aggregated schwertmannite: A novel adsorbent for the efficient removal of arsenic. *Journal of Water Resources Engineering* 3: 1–9.

Dixit S and Hering JG (2003). Comparison of arsenic(V) and arsenic(III) sorption onto iron oxide minerals: Implications for arsenic mobility. *Environmental Science & Technology* 37: 4182–4189.

Dong S, Dou X, Mohan D, Pittman Jr. CU and Luo J (2015). Synthesis of graphene oxide/schwertmannite nanocomposites and their application in Sb(V) adsorption from water. *Chemical Engineering Journal* 270: 205–214.

Dou X, Mohan D and Pittman Jr. CU (2013). Arsenate adsorption on three types of granular schwertmannite. *Water R search* 47: 2938–2948.

Duan H, Liu Y, Yin X, Bai J and Qi J (2016). Degradation of nitrobenzene by Fenton-like reaction in a H_2O_2/schwertmannite system. *Chemical Engineering Journal* 283: 873–879.

Duker AA, Carranza EJM and Hale M (2005). Arsenic geochemistry and health. *Environment International* 31: 631–641.

Egal M, Casiot C, Morin G, Parmentier M, Bruneel O, Lebrun S and Elbaz-Poulichet F (2009). Kinetic control on the formation of tooeleite, schwertmannite and jarosite by *Acidithiobacillus ferrooxidans* strains in an As(III)-rich acid mine water. *Chemical Geology* 265: 432–441.

Eskandarpour A, Onyango MS, Ochieng A and Asai S (2008). Removal of fluoride ions from aqueous solution at low pH using schwertmannite. *Journal of Hazardous Materials* 152: 571–579.

Eskandarpour A, Sassa K, Bando Y, Okido M and Asai S (2006). Magnetic removal of phosphate from wastewater using schwertmannite. *Materials Transactions* 47: 1832–1837.

Fan C, Guo C, Zeng Y, Tu Z, Ji Y, Reinfelder JR, Chen M, Huang W, Lu G, Yi X and Dang Z (2019). The behavior of chromium and arsenic associated with redox transformation of schwertmannite in AMD environment. *Chemosphere* 222: 945–953.

Fan L and Zhang X (2017). Adsorption and desorption of cadmium on synthetic schwertmannite. *Desalination and Water Treatment* 79: 243–250.

Fernandez-Martinez A, Timon V, Roman-Ross G, Cuello GJ, Daniels JE and Ayora C (2010). The structure of schwertmannite, a nanocrystalline iron oxyhydroxysulfate. *American Mineralogist* 95: 1312–1322.

Fukushi K, Sasaki M, Sato T, Yanase N, Amado H and Ikeda H (2003a). A natural attenuation of arsenic in drainage from an abandoned arsenic mine dump. *Applied Geochemistry* 18: 1267–1278.

Fukushi K, Sato T and Yanase N (2003b). Solid-solution reactions in As(V) sorption by schwertmannite. *Environmental Science & Technology* 37: 3581–3586.

Gan M, Zheng Z, Sun S, Zhu J and Liu X (2015). The influence of aluminum chloride on biosynthetic schwertmannite and Cu(II)/Cr(VI) adsorption. *RSC Advances* 5: 94500.

Goswami A and Purkait MK (2014). Removal of fluoride from drinking water using nanomagnetite aggregated schwertmannite. *Journal of Water Process Engineering* 1: 91–100.

Guo J, Dong C, Zhang J and Lan Y (2015). Biogenic synthetic schwertmannite photocatalytic degradation of acid orange 7 (AO7) assisted by citric acid. *Separation and Purification Technology* 143: 27–31.

Guo X, Wu Z and He M (2009). Removal of antimony(V) and antimony(III) from drinking water by coagulation–flocculation–sedimentation (CFS). *Water Research* 43: 4327–4335.

HoungAloune S, Hiroyoshi N and Ito M (2017). Effect of reaction temperature on schwertmannite synthesis, from simulated copper heap leach solutions. *International Journal of Management Information Technology and Engineering* 5: 27–34.

HoungAloune S, Kawaai T, Hiroyoshi N and Ito M (2014). Study on schwertmannite production from copper heap leach solutions and its efficiency in arsenic removal from acidic sulfate solutions. *Hydrometallurgy* 147–148: 30–40.

Hu X, Peng X and Kong L (2017). Removal of fluoride from zinc sulfate solution by in situ Fe(III) in a cleaner desulfuration process. *Journal of Cleaner Production* 164: 163–170.

Jiang D, Li Y, Wu Y, Zhou P, Lan Y and Zhou L (2012). Photocatalytic reduction of Cr(VI) by small molecular weight organic acids over schwertmannite. *Chemosphere* 89: 832–837.

Johnston SG, Keene AF, B urton ED, Bush RT, Sullivan LA, Mcelnea AE, Ahern CR, Smith CD, Powel B and Hocking RK (2010). Arsenic mobilization in a seawater inundated acid sulfate soil. *Environmental Science & Technology* 44: 1968–1973.

Jönsson J, Persson P, Sjoberg S and Lovgren L (2005). Schwertmannite precipitated from acid mine drainage: Phase transformation, sulphate release and surface properties. *Applied Geochemistry* 20: 179–191.

Khamphila K, Kodama R, Sato T and Otake T (2017). Adsorption and post adsorption behavior of schwertmannite with various oxyanions. *Journal of Minerals and Materials Characterization Engineering* 5: 90–106.

Kinsela AS, Collins RN and Waite TV (2011). Speciation and transport of arsenic in an acid sulfate soil-dominated catchment, eastern Australia. *Chemosphere* 82: 879–887.

Kumpulainen S, Räisänen ML, Von der Kammer F and Hofmann T (2008). Ageing of synthetic and natural schwertmannites at pH 2–8. *Clay Minerals* 43: 437–448.

Lan YQ, Yang JX and Deng B (2006). Catalysis of dissolved and adsorbed iron(II) in soil suspension on chromium(VI) reduction by sulfide. *Pedosphere* 16: 572–578.

Liao Y, Liang J and Zhou L (2011). Adsorptive removal of As(III) by biogenic schwertmannite from simulated As-contaminated groundwater. *Chemosphere* 83: 295–301.

Lozano A, Ayora C and Fernández-Martínez A (2020). Sorption of rare earth elements on schwertmannite and their mobility in acid mine drainage treatments. *Applied Geochemistry* 113: 104499.

Maillot F, Morin G, Juillot F, Bruneel O, Casiot C, Ona-Nguema G, Wang Y, Lebrun S, Aubry E, Vlaic G and Brown Jr. GE (2013). Structure and reactivity of As(III)- and As(V)-richschwertmannites and amorphous ferric arsenate sulfatefrom the Carnoule's acid mine drainage, France: Comparisonwith biotic and abiotic model compoundsand implications for As remediation. *Geochimica et Cosmochimica Acta* 104: 310–329.

Mandal BK and Suzuki KT (2002). Arsenic round the world: A review. *Talanta* 58: 201–235.

Marescotti P, Carbone C, Comodi P, Frondini F and Lucchetti G (2012). Mineralogical and chemical evolution of ochreous precipitates from the Libiola Fe–Cu-sulfide mine (Eastern Liguria, Italy). *Applied Geochemistry* 27: 577–589.

Marouane B, Klug M, As K, Engel J, Reichel S, Janneck E and Peiffer S (2021). The potential of granulated schwertmannite adsorbents to remove oxyanions (SeO_3^{2-}, SeO42–, MoO42–, PO43–, Sb(OH)6–) from contaminated water. *Journal of Geochemical Exploration* 223: 106708.

Mohan D and Pittman CU (2007). Arsenic removal from water/wastewaterusing adsorbents—A critical review. *Journal of Hazardous Materials* 42: 1–53.

Mohapatra M, Anand S, Mishra BK, Giles DE and Singh P (2009). Review of fluoride removal from drinking water. *Journal of Environmental Management* 91: 67–77.

Morin G, Juillot F, Casiot C, Bruneel O, Personné J, Elbaz-Poulichet F, Leblang M, Ildefonse P and Calas G (2003). Bacterial formation of tooeleite and mixed arsenic(III) or arsenic(V)-iron(III) gels in the Carnoule's acid mine drainage, France. A XANES, XRD, and SEM study. *Environmental Science & Technology* 37: 1705–1712.

Münch M, van Kaam R, As K, Peiffer S, ter Heerdt G, Slomp C and Behrends T (2024). Impact of ion addition on phosphorus dynamics in sediments of shallow peat lake 10 years after treatment. *Water Research* 248: 120844.

Ooi K, Sonoda A, Makita Y and Torimura M (2017). Comparative study on phosphate adsorption by inorganic and organic adsorbents from a diluted solution. *Journal of Environmental Chemical Engineering* 5: 3181–3189.

Paikaray S, Essilfie-Dughan J, Göttlicher J, Pollock K and Peiffer S (2014). Redox stability of As(III) on schwertmannite surfaces. *Journal of Hazardous Materials* 265: 208–216.

Paikaray S, Göttlicher J and Peiffer S (2011). Removal of As(III) from acidic waters using schwertmannite: Surface speciation and effect of synthesis pathway. *Chemical Geology* 283: 134–142.

Park JH, Han Y and Ahn JS (2016). Comparison of arsenic co-precipitation and adsorption by ironminerals and the mechanism of arsenic natural attenuation in a minestream. *Water Research* 106: 295–303.

Pierce ML and Moore CB (1982). Adsorption of arsenite and arsenate on amorphous iron hydroxide. *Water Research* 16: 1247–1253.

Regenspurg S and Peiffer S (2005). Arsenate and chromate incorporationin schwertmannite. *Applied Geochemistry* 20: 1226–1239.

Reimer KJ and Thompson JAJ (1988). Arsenic speciation in marine interstitial water. The occurrence of organoarsenicals. *Biogeochemistry* 6: 211–237.

Schoepfer VA and Burton ED (2021). Schwertmannite: A review of its occurrence, formation, structure, stability and interactions with oxyanions. *Earth-Science Reviews* 221: 103811.

Sidenko NV and Sherriff BL (2005). The attenuation of Ni, Zn and Cu, by secondary Fe phases of different crystallinity from surface and ground water of two sulfide mine tailings in Manitoba, Canada. *Applied Geochemistry* 20: 1180–1194.

Smedley PL and Kinniburgh DG (2002). A review of the source, behaviour and distribution of arsenic in natural waters. *Applied Geochemistry* 17: 517–568.

Song J, Jia SY, Ren HT, Wu SH and Han X (2015). Application of a high-surface-area schwertmannite in the removal of arsenate and arsenite. *International Journal of Environmental Science and Technology* 12: 1559–1568.

Sthiannopkao S, Kim KW, Cho KH, Wantala K, Sotham S, Sokuntheara C and Kim JH (2010). Arsenic levels in human hair, Kandal Province, Cambodia: The influences of groundwater arsenic, consumption period, age and gender. *Applied Geochemistry* 25: 81–90.

Stumm W and Morgan J (1995). *Aquatic chemistry: Chemical equilibria and rates in natural waters*. 3rd Ed. Wiley. ISBN: 978-0-471-51185-4.

Swedlund PJ and Webster JG (2001). Cu and Zn ternary surface complex formation with SO_4 on ferrihydrite and schwertmannite. *Applied Geochemistry* 16: 503–511.

USEPA (1984). *Antimony: An environmental and health effects assessment*. US Environmental Protection Agency, Office of Drinking Water.

Wan J, Guo C, Tu Z Zeng Y, Fan C, Lu G and Dang Z (2018). Microbial reduction of Cr(VI)-loaded schwertmannite by *Shewanellaoneidensis* MR-1. *Geomicrobiology Journal* 35: 727–734.

Wang R, Guo Y, Song Y, Guo Y, Wang X, Yuan Q, Ning Z, Liu C, Zhou L and Zheng G (2023). Remediating flooding paddy soils with schwertmannite greatly reduced arsenic accumulation in rice (Oryza sativa L.) but did not decrease the utilization efficiency of P fertilizer. *Environmental Pollution* 324: 121383.

Wang W, Song J and Han X (2013). Schwertmannite as a new Fenton-like catalyst in the oxidation of phenol by H_2O_2. *Journal of Hazardous Materials* 262: 412–419.

Wang X, Jiang H, Fang D, Liang J and Zhou L (2019). A novel approach to rapidly purify acid mine drainage through chemically forming schwertmannite followed by lime neutralization. *Water Research* 151: 515–522.

Wang Y, Gao M, Huang W, Wang T and Liu Y (2020). Effects of extreme pH conditions on the stability of As(V)-bearing schwertmannite. *Chemosphere* 251: 126427.

Waychunas GA, Xu N, Fuller CC, Davis JA and Bigham JM (1995). XAS study of AsO_4^{3-} and SeO_4^{2-} substituted schwertmannites. *Physica B* 208–209: 481–483.

Webster JG, Swedlund PJ and Webster KS (1998). Trace metal adsorption onto a acid mine drainage iron(III) oxy hydroxy sulfate. *Environmental Science & Technology* 32: 1361–1368.

Wei X, Viadero RC and Bhojappa S (2008). Phosphorus removal by acid mine drainage sludge from secondary effluents of municipal wastewater treatment plants. *Water Research* 42: 3275–3284.

Wilkin RT, Wallschlager D and Ford RG (2003). Speciation of arsenic in sulfidic waters. *Geochemical Transactions* 4: 1–7.

Wu Y, Guo J, Jiang D, Zhou P, Lan Y and Zhou L (2012). Heterogeneous photocatalytic degradation of methyl orange in schwertmannite/oxalate suspension under UV irradiation. *Environmental Science and Pollution Research* 19: 2313–2320.

Xiong H, Hu D, Shi K, Zhu S and Xu W (2023). Adsorptive removal of arsenic ions from contaminated water using low-cost schwertmannites and akaganéites. *Materials Chemistry and Physics* 297: 127411.

Yan S, Zheng G, Meng X and Zhou L (2017). Assessment of catalytic activities of selected iron hydroxysulphates biosynthesized using Acidithiobacillusferrooxidans for the degradation of phenol in heterogeneous Fenton-like reactions. *Separation and Purification Technology* 185: 83–93.

Yang Z, Wu Z, Liao Y, Liao Q, Yang W and Chai L (2017). Combination of microbial oxidation and biogenic schwertmannite immobilization: A potential remediation for highly arsenic contaminated soil. *Chemosphere* 181: 1–8.

Yu C, Zhang J, Wu X, Lan Y and Zhou L (2014). Cr(VI) removal by biogenic schwertmannite in continuous flow column. *Geochemical Journal* 48: 1–7.

Yu JY and Heo B (2001). Dilution and removal of dissolved metals from acid mine drainage along Imgok creek, Korea. *Applied Geochemistry* 16: 1041–1053.

Yue X and Li-Xiang Z (2013). Arsenite removal from simulated groundwater by biogenic schwertmannite: A column trial. *Pedosphere* 23: 402–408.

Zhang S, Jia S, Yu Bo, Liu Y, Wu S and Han X (2016). Sulfidization of As(V)-containing schwertmannite and its impact on arsenic mobilization. *Chemical Geology* 420: 270–279.

Zhang Z, Guo G, Li X, Zhao Q, Bi X, Wu K and Chen H (2019). Effects of hydrogen-peroxide supply rate on schwertmannite microstructure and chromium(VI) adsorption performance. *Journal of Hazardous Materials* 367: 520–528.

Zhao H, Xia B, Qin J and Zhang J (2012). Hydrogeochemical and mineralogical characteristics related to heavy metal attenuation in a stream polluted by acid mine drainage: A case study in Dabaoshan Mine, China. *Journal of Environmental Sciences* 24: 979–989.

Zhou P, Li Y, Shen Y, Lan Y and Zhou L (2012). Facilitating role of biogenetic schwertmannite in the reduction of Cr(VI) by sulfide and its mechanism. *Journal of Hazardous Materials* 237–238: 194–198.

Zhu F, Guo Z and Hu X (2020). Fluoride removal efficiencies and mechanism of schwertmannite from $KMnO_4/MnO_2$–Fe(II) processes. *Journal of Hazardous Materials* 397: 122789.

Zhu J, Gan M, Zhang D, Hu Y and Chai L (2013). The nature of schwertmannite and jarosite mediated by two strains of Acidithiobacillus ferrooxidans with different ferrous oxidation ability. *Materials Science and Engineering C* 33: 2679–2685.

6

STABILITY OF SCHWERTMANNITE IN THE ENVIRONMENT

6.1 Introduction

Schwertmannite is a metastable phase, its stability being related to biogeochemical processes that interfere with its main components Fe(III) and sulphate. The commonly observed end products of schwertmannite transformation are goethite, lepidocrocite, and hematite often forming through an intermediate ferrihydrite phase under suitable conditions. Under acidic conditions also jarosite may form. These transformation products undergo progressive structural modifications to achieve the most thermodynamically stable state. For instance, ferrihydrite initially forms and subsequently transforms into goethite, while goethite itself may further convert into hematite, depending on the specific environmental conditions affecting schwertmannite exposure. Two primary mechanisms govern the transformation of schwertmannite into stable crystalline end products, which are dissolution–re-precipitation and solid-state transformation. While solid-state transformation is primarily related to higher-temperature processes, dissolution of schwertmannite followed by re-precipitation of a new ferric phase essentially occurs under environmental conditions. Altogether, the following major alteration reactions involved in the transformation of schwertmannite need to be considered:

- Transformation to more stable ferric minerals both under alkaline or acidic conditions
- Interference with Fe(II) to undergo Fe(II)-catalyzed transformation in Fe(II)-bearing (hence anoxic) environments

DOI: 10.1201/9781003583875-6

- Removal of SO_4^{2-}, due to exchange with other anions or by dilution of SO_4^{2-}-rich waters
- Microbial reduction of both Fe(III) and SO_4^{2-}
- Temperature driven alterations

In this chapter we will discuss the key processes driving these reactions. In addition, we will discuss the extent of release of ions, in particularly Fe^{3+} and SO_4^{2-}, but also other ions and the processes controlling the release.

6.2 Transformation to more stable ferric minerals

The metastability of schwertmannite is mainly driven by the competition between SO_4^{2-} and OH^- for Fe^{3+} to form a solid precipitate. Its transformation to goethite is an acidity-generating dissolution and re-precipitation (dissolution–re-precipitation) process (Eq. 6.1).

$$\begin{aligned} 2xH_2O + Fe_8O_8(OH)_{8-2x}(SO_4)_x &\rightarrow \\ 8FeOOH + xSO_4^{2-} + 2xH^+ \ (1 \leq x \leq 1.75) \end{aligned} \quad \text{(Eq. 6.1)}$$

As a consequence, the rate of transformation is strongly depending on pH which is equivalent with an increasing activity of OH-ions.

Hence, under acidic conditions, the complete transformation of schwertmannite in oxygen-rich environments typically requires months to years, or even decades, with goethite emerging as the most stable end product (Bigham et al., 1996a; Regenspurg et al., 2004; Paikaray and Peiffer, 2012; Chen et al., 2022). A pH drop through schwertmannite transformation following Eq. 6.1 has been documented widely (Acero et al., 2006; Bigham et al., 1996a). Murad and Rojik (2003, 2004) report stability and predominance of schwertmannite with traces of goethite consecutively for 2 years (May 2002–July 2004) in a stream bed affected by acid mine drainage (AMD) under extremely low freshwater stream flows following the hot and dry summer of 2003 and thereby maintaining a streamflow pH of 3.0–3.3 during this period. Under conditions of dilute mine waters with comparatively low SO_4^{2-} concentrations, e.g., in streams affected by AMD inflows, co-occurrence of both minerals is possible (Brady et al., 1986).

Overall, once formed, schwertmannite acts as a pH buffer system establishing a characteristic pH stability field in diluted acid mine water affected systems like mining pit lakes or streams around pH 3 (Regenspurg et al., 2004).

Goethite and jarosite tend to form under highly acidic conditions ($pH < 2.5$) (Fukushi et al., 2003; Gagliano et al., 2004; Jönsson et al., 2005, 2006; Acero et al., 2005, 2006). The presence of cations, such as K^+, NH_4^+, or H_3O^+, is favouring jarosite formation and facilitates transformation by incorporating these cations into the jarosite structure under highly acidic and sulphate-rich mine waters, where schwertmannite is thermodynamically unstable (Barham, 1997; Kumpulainen et al., 2008). Wang et al. (2006) demonstrated that the Fe^{2+}-oxidizing bacteria *Acidithiobacillus ferrooxidans* initially generated schwertmannite from oxidation of Fe(II) at pH 2. In the presence of high concentrations of NH_4^+ (100 mM), jarosite formation was induced within 21 days with the rate of transformation increasing at higher temperatures (Figure 6.1b). At lower NH_4^+ concentrations, schwertmannite mineralogy remained unchanged for the same period and transformation occurred only after three weeks.

Schroth and Parnell (2005) reported goethite formation from schwertmannite at pH < 2.5 near the abandoned Pb–Ag Alta mine in Montana, USA, attributing the pH decrease to the acidity generated during schwertmannite transformation. Similarly, Sánchez-España et al. (2012) observed the presence of H_3O^+-jarosite at depths of up to 100 m in a mine pit lake, where pH varied narrowly between 2.2 and 3.0 (Figure 6.1a). Acero et al. (2006) further documented a pH decline from 3.09 to 1.74 over 148 days of schwertmannite ageing, coinciding with jarosite and goethite formation and the release of Fe^{3+} and SO_4^{2-}. Goethite frequently coexists with jarosite under acidic conditions due to its broader stability range (Antelo et al., 2013). Goethite can also form from schwertmannite after prolonged exposure to acidic sediment pore water (pH < 4), with transformations observed over a 12-month period (Vithana et al., 2015).

Vertical profiles of acid sulphate soils (ASSs) reveal a phase transformation pattern in which schwertmannite dominates near the surface (pH < 4), while goethite and jarosite are more prevalent a few centimetres below (pH < 3.5, Acero et al., 2006; Peretyazhko et al., 2009; Kim and Kim, 2021). This results in an increasing goethite-to-schwertmannite ratio with depth in ochreous sediments in a constructed mine drainage wetland (Gagliano et al., 2004). For instance, goethite formation within 5 cm below the surface has been documented at the Dalsung mine, South Korea (Kim and Kim, 2021). Seasonal variations in mineralogy further influence these transformations, with goethite dominating along acidic mine drainage flows during late winter and early spring (February) following snowmelt, whereas schwertmannite predominates at depths of up to 2 m, with goethite becoming more prevalent beyond this depth during summer (May) (Kumpulainen et al., 2007; Peretyazhko et al., 2009).

Murad and Rojik (2003) observed the occurrence of ferrihydrite near the confluence of an acidic mine effluent (pH ~ 3.7) from the Lomnice mine in the Czech Republic with an alkaline stream (pH ~ 8.4), as well as further downstream from the confluence, where the pH stabilizes at approximately 7.3. However, the mine effluent itself is primarily dominated by

FIGURE 6.1 Scanning electron microscope images showing (a) appearance of jarosite (rounded globular) within natural schwertmannite particles found in the acidic pit lake of San Telmo, Huelva, Spain (ref. Sánchez-España et al., 2012) and (b) NH^{4+}/H_3O^+ jarosite formed in *A. ferrooxidans* culture media in presence of 165.4 mM NH_4^+, 37°C within 7 days (ref. Wang et al., 2006).

schwertmannite. Ferrihydrite can form as a transformation product of schwertmannite, and notably, schwertmannite is not observed further downstream of the confluence. The presence of ferrihydrite has also been documented in other mine drainage environments, particularly at pH levels exceeding 5.0 (Winland et al., 1991; Bigham et al., 1992, 1996a; Schwertmann et al., 1995). Additionally, goethite, along with 2-line ferrihydrite, tends to dominate at pH levels above 5.5, as evidenced by both natural observations and laboratory studies on schwertmannite ageing (Winland et al., 1991; Bigham et al., 1996a). This mineralogical progression is supported by multiple studies, which indicate that schwertmannite is most abundant between pH 3 and 4, ferrihydrite predominates at pH 5–8, and goethite occurs within a pH range of 2.5–7.5 (Bigham et al., 1992).

6.2.1 Effect of pH on transformation

The discussion above clearly demonstrates that the transformation product is dependent on pH driving the dissolution–re-precipitation process. Transformation to a stable ferric (oxyhydr)oxide end product is controlled by the rate at which the protons (H^+) in Eq. 6.1 are neutralized, i.e., by the supply rate of alkalinity. Based on SO_4^{2-} release rates, Regenspurg et al. (2004) determined a rate dependency of the schwertmannite transformation rate R_{trnf} on pH to be (Eq. 6.2)

$$R_{trnf} = 0.28pH - 0.80 \qquad \text{(Eq. 6.2)}$$

Transformation to goethite has been demonstrated to occur in the sediment of an acidic mining lake at a depth of ~10 cm where the pH of the pore water increased from <3.5 to >6 (Peine et al., 2000) at a rate of 3.5 mol m^{-2} a^{-1}. The reaction was driven by net alkalinity-generating reactions in the sediment through anaerobic alkalinity-generating microbial processes.

No mineralogical changes are normally seen at pH ≤ 6.0 with the first days of exposure to these conditions (Paikaray and Peiffer, 2012; Li et al., 2018), while clear signs of transformation to goethite were visible thereafter. Similarly, initial schwertmannite transformation to poorly crystalline ferric precipitates resembling ferrihydrite was observed at pH 5.5 (Jönsson et al., 2006), which then partly transformed to goethite and lepidocrocite after 17 days. At pH 7 lepidocrocite has formed. When the pH was increased to 9, goethite formed already after 24 hours (Jönsson et al., 2005). In contrast, Schwertmann and Carlson (2005) found goethite after 79 days of ageing of pure schwertmannite at pH 7 in sulphate-free water, while it took almost 100 days at pH 6. These differences in observation may be due not only to differences in pH, but also to the composition of the solution, i.e., the absence or presence of SO_4^{2-} in which the experiments were performed. Dilute water drives desorption of sulphate which then triggers and accelerates the dissolution–re-precipitation process.

Structural investigations of the ageing process using Fourier transform infrared spectroscopy and Mössbauer spectroscopy revealed an initial phase of transformation ("proto transformation") of schwertmannite. Such phase is characterized by a loss in characteristic Fe–SO_4 domains with increasing pH (As et al., 2024). Mössbauer signals exhibited an increase of the ratio between collapsed sextets (Fe–O bonds) to doublets (Fe–SO_4 domains) at pH 8 after 168 hours of reaction. Despite these structural changes, no substantial morphological changes could be observed and no changes in X-ray diffraction (XRD) patterns.

6.2.2 *Inhibitory effects*

Due to its high specific surface area (SSA) and poor crystallinity, schwertmannite is an effective scavenger of trace elements (e.g., As, Zn, Mo, Cu) from mine effluents (Waychunas et al., 1995; Webster et al., 1998; Randall et al., 1999; Swedlund and Webster, 2001; Carlson et al., 2002; Fukushi et al., 2003; Walter et al., 2003; Jönsson et al., 2005; Courtin-Nomade et al., 2005) and a potential adsorbent for wastewater treatment (Regenspurg and Peiffer, 2005; Burton et al., 2010; Paikaray et al., 2011, 2012, 2014; Janneck et al., 2011, 2015; Klug et al., 2016). The uptake of these contaminants is governed by both structural incorporation and ligand exchange with surface-bound SO_4^{2-} (Fukushi et al., 2004; Jönsson et al., 2005; Burton et al., 2010; Paikaray et al., 2011, 2012). This strong retention affinity has been shown to enhance schwertmannite stability in the presence of As(III) (Paikaray et al., 2011), Cu(II) (Antelo et al., 2013; Li et al., 2018), As(V), Cr(VI), Se(VI) (Liao et al., 2011; Cruz-Hermández et al., 2017), Co(II) (Schonberger, 2016), or oxyanions like PO_4^{3-} (Liao et al., 2011, Schoepfer et al., 2017; As et al., 2024).

The solubility of schwertmannite is decreasing upon contaminant uptake. The binding of oxyanions strengthens Fe(III)–metal interactions within schwertmannite, similar to Fe(III)–SO_4 bonding, the inhibiting effect increasing with the affinity of the ion and amount of ions adsorbed (Antelo et al., 2013; Houngaloune et al., 2015b; Khamphila et al., 2017; Schoepfer et al., 2017). As a consequence, the release of Fe^{3+} and SO_4^{2-} is reduced (Antelo et al., 2013; Schonberger, 2016; Khamphila et al., 2017), preventing the precipitation of new ferric oxyhydroxide phases. Schwertmannite has shown a strong affinity for PO_4^{3-} even up to ~650 µmol g^{-1} (Schoepfer et al., 2017) that inhibits goethite formation marginally (Schoepfer et al., 2017). Goethite formation was observed after 5 days of ageing of pure schwertmannite, but 80 μmol g^{-1} PO_4^{3-} loading delayed its formation until 9 days and no goethite formed until 41 days in >400 μmol g^{-1} PO_4^{3-} loaded schwertmannite. In this study, Fe(II) has accumulated in the solution due to dissimilatory reduction of schwertmannite-Fe(III) in reducing condition.

The higher affinity of As(V) and Mo(VI) towards schwertmannite compared to Cr(VI) enhanced stabilization of schwertmannite (Li et al., 2018), which, in turn, was directly correlated with the fraction of incorporated oxyanions. Similarly, Wang et al. (2020) found that As(V)-rich schwertmannite (containing 2.9% and 9% As(V)) substantially delayed the dissolution–transformation process. Only trace amounts of goethite were found after 30 days at extreme pH conditions (pH 2 and pH 10).

Moreover, the stability of schwertmannite appears to be also controlled by the way in which an ion is taken up. Schwertmannite to which As(V) was adsorbed had a higher stability when compared to specimen into which As(V) was incorporated through co-precipitation with Fe^3 and SO_4^{2-}, as indicated by the slightly larger release rate of sulphate from the co-precipitate (Wang et al., 2020).

Schoepfer et al. (2019) reported the formation of a microcrystalline Fe(III)–oxyhydroxide phase upon the interaction of schwertmannite with PO_4^{3-} at loadings of 80, 400, and 800 μmol g^{-1}. This phase was XRD amorphous due to its small crystallite size and lack of long-range order. Because PO_4^{3-} is larger than SO_4^{2-}, its exchange with SO_4^{2-} in schwertmannite is regarded to induce structural distortion, ultimately leading to the collapse of the schwertmannite tunnel structure. This transformation is characterized by a collapsed sextet spectral signature as observed in Mössbauer spectroscopy. The proportion of this newly formed phase increases with PO_4^{3-} content, comprising 25%, 36%, and 58% of the total iron content of the initial schwertmannite at PO_4^{3-} loadings of 80, 400, and 800 μmol g^{-1}, respectively.

The presence of natural organic matter (NOM) may substantially delay the transformation of schwertmannite (Kumpulainen et al., 2008) in oxic aqueous environments. It negatively impacts product formation while slightly enhancing schwertmannite stability (Jönsson et al., 2006). NOM reduces the release rate of sulphate (SO_4^{2-}) from schwertmannite, thereby lowering its dissolution rate and contributing to its overall stability (Knorr and Blodau, 2007). This effect is primarily due to adsorption of NOM onto schwertmannite surfaces, which hinders dissolution and further stabilizes the mineral (Knorr and Blodau, 2007).

Interestingly, also the sulphate content may exert an inhibitory effect. If the initial SO_4^{2-} content of schwertmannite is high, then it may decelerate its dissolution and thereby limit the availability of Fe^{3+} in solution for the formation of Fe–O oligomers that may then further condense to a new solid phase. Furthermore, once released, dissolved SO_4^{2-} may interfere with these oligomers and restrict nucleation. The first mechanism more effectively explains the slow transformation rate, as SO_4^{2-} adsorption under alkaline pH conditions is relatively low (Davidson et al., 2008). In fact, goethite crystallization rates from pure ferrihydrite were much faster than those from schwertmannite (Davidson et al., 2008).

Schwertmannite transformation may be also inhibited by the concentration of dissolved SO_4^{2-} that reduces the activity of Fe^{3+} ions in solution and thus limits the availability of Fe^{3+} for the formation of Fe–O oligomers that may then further condense to a new solid phase. Furthermore, once released, dissolved SO_4^{2-} may interfere with these oligomers and restrict nucleation. Burton et al. (2013) reported that in sulphate-reducing (SRB) and fermentative (FRB) bacterial systems, schwertmannite completely transformed into mackinawite (~30%) and goethite (~65%) after 17 days, and into mackinawite (~30%) and siderite (~65%) after 400 days. However, in the presence of 100 mM dissolved SO_4^{2-}, approximately 30% of the schwertmannite remained untransformed after 400 days, coexisting with mackinawite (~30%) and siderite (~40%).

6.3 Microbial reduction of schwertmannite

The development of anoxic conditions is common in pit-mine lake environments (Peine et al., 2000; Blodau, 2004) and ASS wetlands (Sullivan and Bush, 2004; Burton et al., 2006). These conditions typically arise due to the influx of large volumes of organic matter (OM) and the re-flooding of constructed wetlands. In acidic drainage, dissolved oxygen (DO) is rapidly depleted because of the high concentrations of Fe^{2+} (10^2–10^3 mg L^{-1}), which readily reacts with available O_2 to form Fe^{3+}, leading to the subsequent precipitation of schwertmannite (Peine et al., 2000; Burton et al., 2007). Consequently, oxygen-limited environments, such as sediments, create favourable conditions for the establishment of reducing conditions. Schwertmannite precipitated in these settings responds dynamically to geochemical changes, influencing ion mobilization and product formation rates.

Anoxic, reducing environments are often characterized by elevated aqueous Fe(II) ($Fe(II)_{aq}$) concentrations, primarily due to microbial activity that facilitates dissimilatory Fe(III) reduction to $Fe(II)_{aq}$ and generates bicarbonate (HCO_3^-), ultimately shifting the pH towards alkaline conditions (Peine et al., 2000; Burton et al., 2008b; Collins et al., 2010). As a result, the reductive dissolution of schwertmannite in anoxic, microbially mediated environments is both an acid-consuming and alkalinity-generating process (Eq. 6.3).

$$3Fe_8O_8(OH)_6SO_4 + C_6H_{12}O_6 + 6H_2O \rightarrow 24Fe^{2+} + SO_4^{2-} + 6CO_2 + 42OH^- \quad \text{(Eq. 6.3)}$$

Both an increase in pH and the presence of dissolved $Fe(II)_{aq}$ contribute to schwertmannite destabilization in anoxic environments. Over time scales ranging from days to years, circum-neutral or alkaline pH alone has minimal effects on schwertmannite stability unless accompanied by elevated $Fe(II)_{aq}$ concentrations. Conversely, schwertmannite remains largely unchanged in acidic conditions, even in the presence of $Fe(II)_{aq}$.

Iron- and sulphur-reducing bacteria are common in mine drainage environments, playing important roles in generating alkalinity through reductive dissolution of ferric (oxy)hydroxides and sulphate (Peine et al., 2000; Pállová et al., 2010; Ke et al., 2024). Ferrous iron produced by reductive dissolution (Eq. 6.3) ultimately yields various Fe–hydroxides/hydroxysulphates as secondary precipitates. Jones et al. (2006) observed differences in Fe(III) reduction efficiency among specific bacterial species. For instance, *Geobacter metallireducens* and *Geobacter* spp. exhibited low efficiency in reducing Fe(III) in schwertmannite at pH 7.0. In contrast, the reduction of jarosite-Fe(III) resulted in the release of high concentrations of Fe^{2+}, SO_4^{2-}, and K^+ into the solution.

In the presence of $Fe(II)_{aq}$, schwertmannite may transform to magnetite upon anaerobic reduction of schwertmannite-Fe(III) by *Geobacter sulfurreducens* in an Fe(II)-catalyzed reaction (Cutting et al., 2012). Reduction of As(V)-loaded schwertmannite sulphate by the SRB *Desulfosporosinus meridiei* produced Mackinawite (FeS), vivianite ($Fe_3(PO_4)_2 \cdot 8H_2O$), and parasymplesite ($Fe_3(AsO_4)_2 \cdot 8H_2O$) (Zhang et al., 2021). The generated sulphide may further react with schwertmannite Fe(III) to produce $Fe(II)_{aq}$ and subsequently FeS (Ke et al., 2024). The intermediate products formed during dissimilatory SO_4^{2-} reduction often lead to cell encrustation limiting potential diffusion of nutrients to cells (Ke et al., 2024).

In a field study performed in sediments of an acidic pit lake, schwertmannite was the predominant ferric mineral in the acidic (pH ~ 3) top sediment (Peine et al., 2000). Schwertmannite was reduced microbially by dissimilatory iron-reducing bacteria (DIRB) in the top layers with acidic conditions being maintained by ongoing transformation into goethite. Only when the acidity generation was finished at greater depth, the pH increased and allowed for sulphate reduction to occur. Similar abundance of goethite at depth has been observed in ASS accompanied by the formation of $FeCO_3$, FeS, and FeS_2 (Burton et al., 2006). Such complex mineralogical assemblages of iron carbonate and ferrous sulphides are indicative for dissimilatory reduction of schwertmannite-Fe(III) at neutral pH condition (Burton et al., 2013). Accumulation of Fe(II) and HCO_3^- upon schwertmannite-Fe(III) reduction results in the formation of $FeCO_3$ while dissimilatory SO_4^{2-} reduction liberates H_2S that eventually generates FeS or FeS_2 (Eq. 6.4) (Burton et al., 2007). Although the activity of SRB in AMD-affected systems is believed to be more prominent at pH > 5.5 (Blodau, 2006; Burton et al., 2013), active participation of acid tolerant sulphate-reducing *Desulfosporosinus* spp. in the production of both FeS and Fe_3S_4 from schwertmannite at pH 4.2 has been described (Bertel et al., 2012). Besides reduction of Fe(III) and SO_4^{2-}, also As(V) incorporated into on schwertmannite may become microbially reduced to As(III) that may ultimately trigger orpiment (As_2S_3) formation (Burton et al., 2013; Eq. 6.5).

$$Fe^{2+} + H_2S \rightarrow FeS + 2H^+, \log K = 3.5 \qquad \text{(Eq. 6.4)}$$

$$H_3AsO_3 + \frac{3}{2}H_2S \rightarrow \frac{1}{2}As_2S_3 + 3H_2O, \log K_{orp} = 11.9 \qquad \text{(Eq. 6.5)}$$

6.4 $Fe(II)_{aq}$-catalyzed transformation in Fe(II) bearing (hence anoxic) environments

Under acidic conditions (pH < 5.0), schwertmannite is regarded to be stable for at least a few days in anoxic environments (Burton et al., 2008b). However, once dissolved $Fe(II)_{aq}$ is introduced into an aqueous system, some transformation to goethite is observed at intermediate pH levels (Burton et al., 2008b; Paikaray et al., 2017), and goethite being the key transformation product at alkaline conditions (Burton et al., 2008b).

This phenomenon is well established for low-crystallinity ferric (oxyhydr)oxides and referred to as Fe(II)-catalyzed transformation (e.g., Gorski and Scherer, 2011). The catalytic role of $Fe(II)_{aq}$ is regarded to follow two mechanistic steps: (1) adsorption of Fe^{2+} onto the mineral surfaces, forming a surface complex, and (2) electron transfer between surface-bound Fe^{2+} and structural Fe^{3+}. Hence, the surface-bound Fe^{3+} may serve as a nucleus for the precipitation of new phase. As in other low-crystallinity ferric oxyhydroxides, such as ferrihydrite, this process ultimately leads to the breakdown of the old mineral structure and in case of schwertmannite, the subsequent release of SO_4^{2-} accompanied by the precipitation of a new ferric (oxyhydr)oxide phase.

Upon dissolution, SO_4^{2-} is released rapidly, reaching nearly 100% within a few hours, particularly under Fe(II)-catalyzed transformation. The rate of SO_4^{2-} release correlates with the rate of goethite formation during Fe(II)-catalyzed transformation. At higher Fe(II) concentrations, the

SO_4^{2-} release rate is increasing, which in turn promotes goethite formation at a faster rate compared to lower Fe(II) concentrations (Burton et al., 2008b).

The extent of adsorption of Fe^{2+} surface complexes on schwertmannite increases with pH, whereas at pH < 5.0, Fe^{2+} adsorption is negligible, preventing complex formation (Burton et al., 2008b, Houngaloune et al., 2015a). Hence, the presence of these surface complexes plays a crucial role in schwertmannite transformation. Under acidic conditions, dissolved Fe(II) in schwertmannite-rich aqueous solutions has a negligible influence on transformation but may act as a strong catalyst at pH > 5.0 (Burton et al., 2008b; Paikaray et al., 2017; Fan et al., 2019). The transformation rate, determined by SO_4^{2-} release, depends on the concentration of Fe_{aq} and pH and significantly reduces the time required for schwertmannite to convert into goethite from hundreds of days to just a few hours. For instance, at 10 mM $Fe(II)_{aq}$ and pH 6.5, schwertmannite transformed into goethite (with traces of lepidocrocite) within 3 hours at a rate of 22.7% h^{-1} (Burton et al., 2008b), whereas in the absence of $Fe(II)_{aq}$, the transformation rate was only 0.004%–0.041% h^{-1} under similar pH conditions (Regenspurg et al., 2004; Jönsson et al., 2005; Schwertmann and Carlson, 2005; Burton et al., 2008b). The formation of goethite and lepidocrocite becomes more rapid and results in higher crystallinity as $Fe(II)_{aq}$ concentration increases (Burton et al., 2008b; Paikaray et al., 2017; Fan et al., 2019). Additionally, the catalytic effect of $Fe(II)_{aq}$ is enhanced at elevated temperatures (Houngaloune et al., 2015a). Fan et al. (2018) reported that schwertmannite remained dominant after 72 hours at pH 6.5 in the presence of 0.1 mM $Fe(II)_{aq}$, whereas complete transformation to goethite occurred within 8 hours at 1 mM $Fe(II)_{aq}$. Even in the presence of oxygen, a similar effect has been observed, where $Fe(II)_{aq}$ promoted goethite formation compared to $Fe(II)_{aq}$-free systems (Ying et al., 2020).

Burton et al. (2008b) observed no mineralogical transformation of schwertmannite in the presence of 5 mM $Fe(II)_{aq}$ at pH < 4.5 for up to 8 days. Similarly, Burton and Johnston (2012) reported no mineralogical changes after 126 days of ageing at pH 6.5 in the absence of Fe(II)aq. Paikaray et al. (2017) confirmed that schwertmannite remained the dominant phase after approximately 8.5 days of exposure to anoxic, $Fe(II)_{aq}$-poor media (1 mM) at pH 5, as well as at pH 7.0 with $Fe(II)_{aq}$ < 1.0 mM. In contrast, at higher $Fe(II)_{aq}$ concentrations and pH > 8.0, goethite can form within hours (e.g., Burton et al., 2008a).

Magnetite was formed upon anaerobic reduction of schwertmannite-Fe(III) by *G. sulfurreducens* to Fe(II) (Cutting et al., 2012) in an Fe(II)-catalyzed reaction. At a high As(V) content of schwertmannite, the transformation was incomplete and required a longer reaction time, while only a little delay was observable at low As(V) content. It was argued that As(V) adsorption increases the ratio between surface-bound Fe(III) and Fe(II) with an increasing As(V) loading potentially limiting the Fe(II) content required for magnetite formation. Instead, a surface coating forms, partially inhibiting the electron transfer between adsorbed Fe^{2+} and bulk Fe^{3+}.

6.4.1 *Influence of water constituents on $Fe(II)_{aq}$-catalyzed transformation*

Even at high $Fe(II)_{aq}$ concentrations, schwertmannite mineralogy remains unchanged in the presence of elevated oxyanion concentrations, such as AsO_3^{3-} or CrO_4^{2-} (Burton et al., 2010; Liao et al., 2011; Paikaray and Peiffer, 2015; Paikaray et al., 2017; Fan et al., 2019). However, the inhibitory response depends on the oxyanion. No goethite formation was observed after 30 days of exposing schwertmannite with a Cr(VI) content of 54 mmol per mol Fe to 4 mM $Fe(II)_{aq}$ at pH 6.5 (Fan et al., 2019). In contrast, goethite was detected after only 8 days when schwertmannite with 65 mmol As(V) per mol Fe was exposed to the same $Fe(II)_{aq}$ concentration. The authors propose that this difference is likely due to differences in the binding mechanisms of the individual oxyanions within schwertmannite. Chromate that becomes structurally incorporated has a stronger inhibiting effect to Fe(II) compared to arsenate that adsorbs on the surface only.

Schwertmannites loaded with Cr^{3+} were relatively stable at pH 6.5 under anoxic conditions with traces of ferrihydrite found for schwertmannite loaded with 0.21 wt% Cr^{3+} after 14 days of reaction (Choppala and Burton, 2018). In contrast, 24% of the initial schwertmannite was transformed to ferrihydrite after 14 days of ageing. The presence of $Fe(II)_{aq}$ (1 and 10 mM) accelerated goethite formation in comparison to experiments performed in the absence of $Fe(II)_{aq}$ formation with lepidocrocite being an intermediate phase occurred. Oxyanion adsorption appears to slow down the transformation rate by inhibiting Fe(II) adsorption onto the schwertmannite surface.

Partial transformation of schwertmannite to which PO_4^{3-} was preadsorbed by Fe(II) has been shown by Schoepfer et al. (2017). The study reports goethite formation after 5 days of ageing of pure schwertmannite, but 80 μmol g^{-1} PO_4^{3-} loading delayed its formation until 9 days and no goethite formed until 41 days > 400 μmol g^{-1} PO_4^{3-} loaded schwertmannite. Fe(II) was produced due to reductive dissolution of Fe(III) in these reactors where the least production took place at high PO_4^{3-} reactors (800 μmol g^{-1}) constituting only 1% of the total Fe(III).

Similar to oxoanions, also dissolved NOM tends to adsorb onto schwertmannite surfaces thereby blocking the surface sites of schwertmannite for Fe(II) adsorption and inhibiting Fe(II)-catalyzed dissolution (Jones et al., 2009).

6.4.2 The effect of silica

After oxygen (~47%), silicon (Si) is the second most abundant element in the Earth's crust (~28%) and a ubiquitous dissolved species in natural water. The affinity of Si for schwertmannite is negligible under acidic conditions, preventing its co-precipitation with schwertmannite (Jones et al., 2009; Burton and Johnston, 2012). However, Si efficiently adsorbs onto schwertmannite under circum-neutral to alkaline pH conditions. Burton and Johnston (2012) demonstrated that at pH 3.5, Si uptake by schwertmannite was almost negligible, whereas at pH 6.5, a Langmuir sorption pattern was observed, indicating monolayer surface confinement of Si on schwertmannite.

Studies on the impact of Si on schwertmannite stability suggest that Si inhibits schwertmannite dissolution, thereby limiting its transformation into more stable crystalline phases. X-ray absorption spectroscopy (extended X-ray absorption fine structure spectroscopy) indicates that Si adsorption reduces the number of iron double-corner linkages during Fe(III) hydrolysis, consequently hindering Fe polymerization into new ferric (oxyhydr)oxides (Doelsch et al., 2000). The inhibitory effect of Si was confirmed in a laboratory study, in which Burton and Johnston (2012) demonstrated that the presence of Si in the aqueous phase generally retarded the overall transformation rate of schwertmannite at pH 6.5. However, under microbially reducing conditions, even in the presence of 9.5 mM Si, the catalytic effect of $Fe(II)_{aq}$ remained active. Hence, $Fe(II)_{aq}$-catalyzed transformation would be the only pathway for transformation of schwertmannite exposed to high Si concentrations. The authors emphasized, though, that in their experimental setup, where all components were introduced simultaneously, $Fe(II)_{aq}$ interacted with schwertmannite more rapidly than a passivating surface layer with Si could form.

In contrast, Collins et al. (2010) reported the persistence of schwertmannite even after 100 years of exposure to acid drainage in coastal lowland in northeastern New South Wales (NSW), Australia. In this environment, biological reduction of schwertmannite was limited, despite $Fe(II)_{aq}$ concentrations remaining around ~1 mM, which the authors attributed to a saturation of the schwertmannite sorption sites by dissolved Si supplied from Si-rich groundwater. Precipitates were extremely enriched with Si varying from 6490 to 31,500 mg Si kg^{-1}. Similarly, Jones et al. (2009) observed no new mineral phase formation from schwertmannite at pH ~ 6.5 in the presence of 1 mM Si. These conflicting findings on the transformation of schwertmannite in the presence of Si are attributed to differences in the timing of Si sorption. When Si interacts with schwertmannite at the same time or later than $Fe(II)_{aq}$, it has a minimal impact on Fe(II)-driven catalysis (Burton and Johnston, 2012). However, adsorption of Si prior to $Fe(II)_{aq}$, as observed in Jones et al. (2009) and Collins et al. (2010), leads to the formation of passivating Si surface species, which obstruct Fe(II) sorption and hinder its catalytic effect. Additionally, the Si content in schwertmannite may also regulate the extent of surface coverage with Si, influencing Fe(II) accessibility for catalytic transformation (Collins et al., 2010). Excessive Si loading on schwertmannite surfaces prevents Fe(II) sorption, thereby limiting schwertmannite transformation.

In summary, the effect of Si on schwertmannite stability is discussed to operate through two primary mechanisms. First, Si adsorption blocks reactive surface sites on schwertmannite, restricting Fe(II)–Fe(III) electron transfer, which is essential for catalyzing schwertmannite transformation. Second, Si in either the solid or aqueous phase inhibits the nucleation and growth of more thermodynamically stable crystalline mineral phases (Cornell and Schwertmann, 2003; Pedersen et al., 2005; Jones et al., 2009).

6.5 Thermal stability

Localities favourable for schwertmannite precipitation are occasionally exposed to high-temperature conditions due to wildfires and forest fires, such as the acid sulphate lowlands of Australia (Bradstock et al., 2006; Henderson et al., 2007; Henderson and Sullivan, 2010; Blake et al., 2012). Organic debris, including leaf litter, is highly combustible and can ignite at temperatures below 400°C. During such events, schwertmannite undergoes structural destabilization (Henderson and Sullivan, 2010), promoting the formation of more stable crystalline iron oxides, such as hematite (α-Fe_2O_3) and maghemite (γ-Fe_2O_3), depending on temperature and duration of thermal exposure. The abundance of hematite in the top 5 cm of coastal floodplain wetlands in NSW, Australia, following bushfires provides strong evidence of this thermal transformation process (Henderson et al., 2007).

Hematite (Fe_2O_3) formation rate from schwertmannite around this locality is accelerated by OM that gains crystallinity upon increasing annealing temperature (Henderson et al., 2007, 2008; Henderson and Sullivan, 2010; Reichelt and Bertau, 2015). Under such scenarios, goethite appears first at ≤80°C followed by hematite at ≥90°C along with appearance of ferrihydrite as an intermediate phase (Davidson et al., 2008). On the other hand, schwertmannite dominance until 200°C and hematite formation at ≥400°C that became dominant as temperature increases to

800°C has been shown by Johnston et al. (2016). Goethite and hematite assemblage between 400°C and 600°C has been found after 90 days of incubation in their study, while no goethite was found at 800°C even from beginning. In addition to hematite and goethite, mackinawite and siderite appear at 400°C and 600°C after 150-day incubation of schwertmannite due to reduction of schwertmannite-Fe(III) and SO_4^{2-}. 2L ferrihydrite as an intermediate phase forms at ~300°C before hematite forms (Reichelt and Bertau, 2015).

Current research indicates that temperatures exceeding 600°C facilitate hematite formation, accompanied by the release of SO_3 gases (Yu et al., 2002; Henderson and Sullivan, 2010). Schwertmannite remains stable up to approximately 200°C, undergoes partial transformation at around 400°C, and completely converts to hematite at temperatures of ≥600°C, with crystallinity increasing as temperature rises (Henderson and Sullivan, 2010; Johnston et al., 2016). In OM-rich schwertmannite, the combustion of OM further accelerates pyrolytic oxidation, leading to weight loss and a decrease in total OM content (Henderson and Sullivan, 2010; Qiao et al., 2017). Goethite (α-FeOOH) emerges as an intermediate phase between 200°C and 600°C along with hematite, with the latter becoming the dominant phase over time and ultimately the sole end product at temperatures exceeding 600°C. At 800°C, hematite forms spontaneously (Johnston et al., 2012).

A study by Vithana et al. (2018) examined the thermal behaviour of schwertmannite containing trace elements such as copper (1.1 µmol g^{-1}), antimony (9.2 µmol g^{-1}), and phosphorus (11.0 µmol g^{-1}). These incorporated elements had minimal impact on initial schwertmannite formation or its subsequent mineralogical transformation. Mineralogical changes remained negligible until 300°C, beyond which hematite peaks began to appear, eventually becoming the dominant and sole phase at 800°C. However, an intermediate iron oxide sulphate phase was detected between 300°C and 500°C. The incorporation of trace metals and nutrients occurred through co-precipitation during schwertmannite formation.

The thermal stability of schwertmannite is also influenced by pH. In highly alkaline conditions (pH > 13), goethite forms at temperatures as low as 80°C and becomes the dominant phase within a few hours (Davidson et al., 2008; Hockridge et al., 2009). In these conditions, 2L ferrihydrite remains the primary intermediate phase. At temperatures exceeding 180°C, hematite emerges as the sole transformation product, with goethite and 2L ferrihydrite serving as transient intermediates. Notably, 2L ferrihydrite forms slightly earlier than goethite in the transformation sequence.

In acidic conditions, goethite can also form within a few hours as the temperature surpasses 60°C (Ying et al., 2020). The transformation rate increases with temperature, as demonstrated by Cruz-Hermández et al. (2017), who observed rapid goethite formation when the ageing temperature was raised from 40°C to 85°C. However, the presence of dissolved As(V) at concentrations exceeding 1 mM significantly inhibited the transformation process by stabilizing schwertmannite. The transformation of schwertmannite follows three distinct stages: (1) an induction period with no detectable transformation products, (2) a primary crystallization stage involving the formation of 2L ferrihydrite, goethite, and/or hematite, and (3) a secondary crystallization phase in which 2L ferrihydrite and goethite transition into hematite.

The overall transformation pathway can be summarized in Figure 6.2: upon exposure to temperatures above 200°C, goethite emerges as the first mineral phase. As temperatures increase beyond 400°C, a mixture of goethite, hematite, magnetite, and 2L ferrihydrite forms. At temperatures above 600°C, hematite becomes the dominant phase. Under anoxic conditions, siderite and mackinawite are the primary transformation products.

Under anoxic conditions, the intermediate and final products of schwertmannite transformation primarily consist of Fe(II)-bearing oxide and sulphide phases. Siderite ($FeCO_3$) and mackinawite (FeS) form from schwertmannite at temperatures exceeding 400°C (Figure 6.3) but disappear at 800°C (Johnston et al., 2016). Additionally, maghemite—a mixed-valence Fe(II)–Fe(III) oxide with a cubic spinel structure similar to magnetite—has

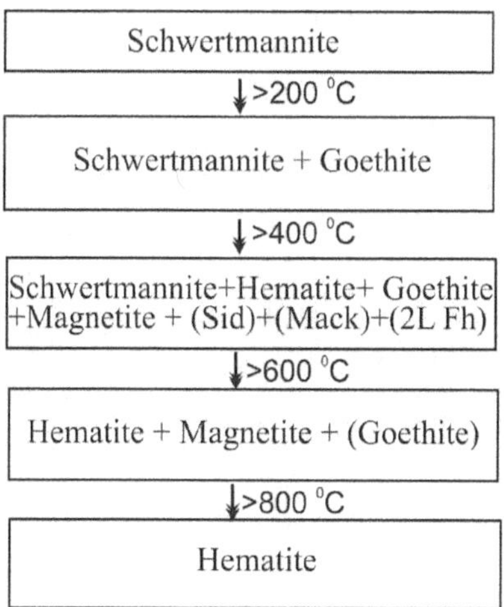

FIGURE 6.2 The overall transformation pathway of schwertmannite at different temperature conditions (Sid = siderite, mack = mackinawite, Fh = ferrihydrite).

TABLE 6.1 End products formed upon schwertmannite transformation at different temperatures under anoxic conditions

Temperature (°C)				
≤200°C	200°C–400°C	400°C–600°C	600°C–800°C	≥800°C
Sch	Sch, Gt, 2L Fh	Sch, Hm, Gt, (Sid), (Mack), 2L Fh, Mht	Hm + Gt + SO_3	Hm

Note: The mineral phases are listed in decreasing abundance, while those within brackets are minor phases. SO_3 is a sulphur gas phase.

Sch = schwertmannite, Gt = goethite, Mht = maghemite, 2L Fh = 2L ferrihydrite, Sid = siderite, Hm = hematite, Mack = mackinawite.

been observed at temperatures above 300°C–400°C (Table 6.1) (Mazzetti and Thistlethwaite, 2002; Johnston et al., 2018).

The generation of Fe(II) occurs through the reduction of Fe(III) within schwertmannite, facilitated by soil OM acting as an electron donor (Johnston et al., 2018). In the absence of OM, heating of schwertmannite exclusively results in the formation of hematite, with no detectable Fe(II) present. In this case, magnetite (Fe_3O_4) serves as an intermediate phase before ultimately transforming into hematite and, in OM-rich environments, maghemite. The highest maghemite yield occurs at approximately 600°C.

Similar to Fe(III) reduction, arsenate (As(V)) incorporated into schwertmannite also undergoes reduction to arsenite (As(III)), with maximum As(III) yield observed at temperatures of up to 400°C.

6.5.1 Mechanisms of transformation at elevated temperature

Ferrihydrite transformation to goethite is commonly attributed to the dissolution–re-precipitation mechanism (Jönsson et al., 2005; Paikaray and Peiffer, 2012). Davidson et al. (2008) reported that hematite can form from aggregates of the ferrihydrite coated with sulphate via this mechanism at approximately 180°C. Such ferrihydrite, which formed as an intermediate from schwertmannite, exhibits a lower solubility due to the presence of sorbed sulphate, which restricts goethite growth and favours hematite formation. However, ferrihydrite aggregates undergo short-term dissolution followed by crystallization within the aggregates, a process strongly dependent on the presence of water (Schwertmann and Murad, 1983; Davidson et al., 2008). An additional factor influencing this transformation is the annealing temperature of 180°C, at which hematite is thermodynamically more stable than goethite (Davidson et al., 2008).

FIGURE 6.3 Scanning electron microscope images showing (a) nano-hematite clusters after thermal transformation of schwertmannite at 800°C and (b) mackinawite, and (c) siderite at 400°C treatment after 150 days of anoxic incubation. Scale bar in each image is 1 micron (ref. Johnston et al., 2016).

The transformation of goethite into hematite at elevated temperatures follows a thermal transformation as documented by Johnston et al. (2016) and Davidson et al. (2008). This process involves the migration of hydroxyls and protons from the structure, with hematite formation progressing at the expense of goethite. The hematite/goethite ratio decreases over time. Structurally, goethite and hematite share similarities, wherein three unit cells of goethite stack along the c-axis to form one unit cell of hematite. At temperatures exceeding 200°C (Henderson and Sullivan, 2010; Johnston et al., 2016; Vithana et al., 2018), a dehydration/dehydroxylation process occurs, accompanied by the release of both surface-bound and structural sulphate. Cation rearrangement follows at temperatures above 500°C (Cudennec and Lecerf, 2005). The release of SO_4^{2-} during dehydration induces structural distortions and defects in schwertmannite (Vithana et al., 2018). The presence of water may trigger this process at lower temperatures (e.g., 150°C; Davidson et al., 2008), although the precise role of water remains unclear.

The point of zero charge (pH_{PZC}) of iron oxides/hydroxides decreases with increasing temperature (Houngaloune et al., 2015a), which may partially explain why Fe^{2+} catalyzes goethite formation at 65°C even under acidic conditions (pH ~ 3.5; Houngaloune et al., 2015a). Adsorbed Cu^{2+} acts as an electron acceptor, reducing electron transfer between Fe^{2+} and Fe^{3+} and thus limiting schwertmannite dissolution, thereby restricting goethite formation.

Annealing at temperatures up to 200°C results in the release of adsorbed water, while structural water is expelled at temperatures exceeding 300°C (Reichelt and Bertau, 2015). This water loss marginally lowers the intensity of infrared (IR) bands corresponding to O–H stretching (~2900–3700 cm^{-1}) and O–H deformation (1635 and 860 cm^{-1}) upon heating to 250°C, with these bands almost disappearing at temperatures beyond 350°C (Qiao et al., 2017). Sulphur dioxide (SO_2) and elemental sulphur (S) are released endothermically beyond 600°C, while schwertmannite transitions to products with decreased water content and increased sulphate content, generating CO_2 between 300°C and 500°C (Henderson and Sullivan, 2010; Reichelt and Bertau, 2015). The intensity of structural sulphate IR bands (981 and 607 cm^{-1}) marginally decreases beyond 550°C. Above 800°C, the end products, which are commonly reported in most studies, contain no residual water or sulphate (Reichelt and Bertau, 2015). At elevated temperatures, particularly beyond 300°C, a hydrogen- and carbon monoxide-rich atmosphere facilitates electron transfer to schwertmannite (Jozwiak et al., 2007; Johnston et al., 2018). This thermally induced electron transfer process promotes the reduction of Fe(III) in schwertmannite, leading to the parallel formation of maghemite and hematite, along with the reduction of schwertmannite-bound contaminants such as arsenic (As(V)) (Johnston et al., 2018).

6.6 Release of ions upon schwertmannite transformation

6.6.1 Schwertmannite Fe^{3+} and SO_4^{2-}

Along with the transformation of schwertmannite, substantial release of species into the solution occurs among which Fe^{3+} and SO_4^{2-} are the most prominent ones. Approximately 35% of the sulphate in schwertmannite is surface bound and can be easily exchanged with other (oxy)anions, making it prone to desorption (Bigham et al., 1990, 1994; Barham, 1997; Paikaray and Peiffer, 2010). Sulphate is the major ion released during the transformation of schwertmannite into goethite or lepidocrocite, under both oxic and anoxic conditions (Regenspurg et al., 2004; Schwertmann and Carlson, 2005; Burton et al., 2008a; Paikaray and Peiffer, 2012, 2015; Paikaray et al., 2017).

In contrast, Fe^{3+} concentrations in the aqueous phase typically remain low, controlled by the dissolution–re-precipitation process (Paikaray and Peiffer, 2012, 2015; Burton et al., 2008b, 2013; Paikaray et al., 2017; Fan et al., 2019). Yet, under acidic conditions, the release of Fe^{3+} is rapid initially but decreases over time (Antelo et al., 2013; Wang et al., 2020). This is due to the initial dissolution of schwertmannite, which facilitates Fe^{3+} release, whereas the following re-precipitation reduces the aqueous Fe^{3+} concentrations again. The rapid release of Fe^{3+} under acidic conditions coincides with an initially slow release of sulphate during the early stage of dissolution, with aqueous Fe^{3+} concentrations decreasing as sulphate release increases in a later stage of the dissolution (Jönsson et al., 2005). This observation suggests that (1) the rate of dissolution exceeds that of re-precipitation during the initial period leading to an accumulation of Fe^{3+}, and (2) schwertmannite dissolution continues during re-precipitation (when dissolved Fe^{3+} concentrations are reduced), leading to continued sulphate release. Thus, both dissolution and re-precipitation processes likely occur simultaneously after a phase of aggressive initial dissolution (Regenspurg et al., 2004; Jönsson et al., 2005; Antelo et al., 2013).

However, a considerable increase in Fe^{3+} concentrations was observed over longer exposure periods, accompanied by a decrease in pH from 2.9 to 1.8 (Acero et al., 2005, 2006; Schonberger, 2016). In a 353-day ageing study, pH decreased to 1.7, Fe^{3+} concentrations exceeded 5600 mg L^{-1}, and SO_4^{2-} concentrations surpassed 18,000 mg L^{-1}, compared to the initial concentrations of ~280 mg L^{-1} of Fe^{3+} and ~3100 mg L^{-1} of SO_4^{2-} (Acero et al., 2006), indicating incomplete or even absent precipitation of Fe, e.g., goethite or jarosite.

The release of sulphate is strongly driven by a shift in the sorption equilibrium triggered by dilution of schwertmannite. Jönsson et al. (2005) observed that the sulphate release increased with decreasing solid-to-water ratio, with a maximum release at low schwertmannite-to-water ratios (0.232 mmol SO_4^{2-} g^{-1} schwertmannite at 1 g L^{-1}, compared to 0.115 mmol SO_4^{2-} g^{-1} at 50 g L^{-1}). Ionic strength only marginally affected the sorption equilibrium below I = 0.1 M, while sulphate release increased by 56% when the ionic strength rose from 0.1 to 2.0 M (Jönsson et al., 2005).

The release of sulphate strongly depends on pH. In a recent paper, As et al. (2024) established a conceptual model for the effect of pH on sulphate release based on own kinetic studies performed after careful pre-equilibration at different pH values and on an extensive literature review.

Depending on pH, up to 60% of the initially present sulphate in schwertmannite was released during the 24-hour pre-equilibration period, consistent with earlier studies (Jönsson et al., 2005). The release of sulphate has primarily been attributed to the transformation to goethite (Bigham et al., 1996a; Murad and Rojík, 2003; Regenspurg et al., 2004; Paikaray and Peiffer, 2010). However, the uncatalyzed transformation of schwertmannite to goethite typically takes much longer (>2 weeks) at room temperature than the 24-hour pre-equilibration period applied in the study by As et al. (2024) (e.g., Burton et al., 2008b). XRD analyses did not indicate any signs of goethite formation. Short-term sulphate release was therefore attributed to the replacement of sulphate groups by hydroxyl groups through a ligand-exchange reaction. Hydroxyl groups replaced sulphate in a stable 2:1 ratio, independent of pH, an observation made earlier by Paikaray and Peiffer (2012) and Jönsson et al. (2005).

Previous X-ray absorption spectroscopy (XAS) analysis identified the inner-sphere sulphate groups in schwertmannite as bidentate–binuclear bridging groups (Wang et al., 2015). Assuming these groups are replaced by two Fe–OH groups (Parfitt and Smart, 1978), the ligand exchange for inner-sphere sulphate was described by Eq. 6.6.

$$\equiv Fe_2 = SO_4 + H_2O \rightarrow FeOH + SO_4^{2-} + 2H^+ \qquad \text{(Eq. 6.6)}$$

where the symbol ≡ denotes a surface group (As et al., 2024). At alkaline pH, the concentration of H^+-ions decreases causing the equilibrium of reaction Eq. 6.6 to shift towards sulphate release. After 24 hours of pre-equilibration, 1.2 mmol SO_4^{2-} g^{-1} was already released at pH 8 (equivalent to 52% of the initial sulphate), which decreased to 0.7 and 0.2 mmol SO_4^{2-} g^{-1} for pH 6 and 3, respectively.

Surface-bound sulphate ions are expected to be preferentially released (Jönsson et al., 2005). Based on mass balances of surface-bound sulphate established in earlier studies, As et al. (2024) conclude that at pH 3 and 6, where sulphate release was below 0.8 mmol SO_4^{2-} g^{-1}, predominantly surface-sulphate, and less tunnel sulphate, was liberated. In contrast, at pH 8, (at least) an additional 0.4 mmol SO_4^{2-} g^{-1} of tunnel sulphate must have been released to attain the equilibrium sulphate release of 1.2 mmol SO_4^{2-} g^{-1}. The existence of two different sulphate pools at alkaline pH aligns with the evaluation of kinetic data from the study by As et al. (2024), which required a two-step model to accurately describe sulphate-release kinetics at pH 8.

Under highly acidic and highly alkaline conditions, sulphate release from schwertmannite can approach 100% (e.g., Burton et al., 2008a). A substantial amount of sulphate is taken up by goethite under acidic conditions, which is the commonly formed end product and which exhibits a strong affinity for surface-adsorbed sulphate (Brady et al., 1986; Bigham et al., 1996b).

Sulphate release is also influenced by the occurrence of $Fe(II)_{aq}$, the effectiveness of which depends on pH. Houngaloune et al. (2015a) observed only little influence of $Fe(II)_{aq}$ at pH 3–4 and T = 25°C even at very high $Fe(II)_{aq}$ concentrations (up to 100 mM). Sulphate release increased from pH 5.5 to 7 at constant Fe^{2+} concentrations ($c(Fe(II)_{aq}$ = 5 mM, Burton et al., 2008b). Paikaray and Peiffer (2015) reported the highest SO_4^{2-} release at pH 8 in the absence of $Fe(II)_{aq}$ compared to that at a concentration of 1 mM $Fe(II)_{aq}$. The authors proposed an inhibition of the dissolution of schwertmannite by adsorption of positively charged Fe(II) species predominating at this pH (Fe^{2+} and $FeOH^{+}$) to the negatively charged surface of schwertmannite and thereby preventing SO_4^{2-} from dissolution. Sulphate release increased with increasing Fe^{2+} concentrations at constant pH (Burton et al, 2008b) and also with temperature even at acidic conditions (T = 65°C, Houngaloune et al., 2015a).

The presence of $Fe(II)_{aq}$ not only affected the sulphate-release rate but also the release pattern changed (Fan et al., 2018). SO_4^{2-} was released instantaneously into a suspension containing 2 g schwertmannite per litre at pH 6.5 in the absence of $Fe(II)_{aq}$ and reached a concentration of 1.9 mM after a few minutes presumably due to desorption and establishing of a new sorption equilibrium. Upon addition of 1 mM $Fe(II)_{aq}$, the SO_4^{2-} concentration increased exponentially reaching a constant concentration of 4.5 mM after 12 hours. The Fe(II)-induced release pattern was modelled by means of a diffusion-controlled process (Fan et al., 2018) and was interpreted as a release of sulphate from the internal structure of schwertmannite upon reduction of structural Fe^{3+} and the splitting of the $Fe–SO_4$ bond. Co-precipitation of Cd^{2+} with schwertmannite only very slightly affected the release rates.

The presence of competing ions reduces release of both Fe^{3+} and SO_4^{2-} under acidic conditions, with minimal release observed at higher ion concentrations (Houngaloune et al., 2015a). Schwertmannite pre-loaded with contaminants such as Co^{2+}, $HAsO_4^{2-}$, CrO_4^{2-}, and MoO_4^{2-} tends to release lower amounts of Fe^{3+} and H^+, likely due to the stabilization of schwertmannite by these sorbed contaminants (Antelo et al., 2013; Li et al., 2018; Schonberger, 2016). In contrast, the presence of Cu^{2+} in the aqueous phase leads to increased Fe^{3+} release, as Cu^{2+} partially substitutes Fe^{3+} in the schwertmannite structure (Schonberger, 2016). Sulphate release also increases in the presence of Cu^{2+} due to the alteration of Fe–SO_4 complexes. Significant sulphate release occurs in the presence of H_3AsO_3 and $HAsO_4^{2-}$ at pH 6.5, with As(V) inducing a greater release compared to As(III) due to the exchange of As(III)/As(V) with sulphate in schwertmannite and the higher affinity of schwertmannite for As(V) (Burton et al., 2010).

Under anoxic conditions, sulphate concentrations decrease after a certain period of ageing, leading to the formation of mackinawite via dissimilatory sulphate reduction by SRBs, which generates H_2S and bicarbonate. Sulphate reduction is thermodynamically favourable at pH > 5 (Blodau, 2006; Burton et al., 2013), and the generated H_2S promotes the formation of mackinawite (Burton et al., 2013).

Sulphate release also affects the SSA of schwertmannite. The SSA of schwertmannite showed an initial increase from 42.9 to >180 $m^2\ g^{-1}$ after 24 hours of ageing at pH 3, and to even 225 $m^2\ g^{-1}$ at pH 7 (Jönsson et al., 2005). This rapid increase in SSA was likely caused by the release of sulphate, which diminished particle agglomeration, thereby increasing surface area available for N_2 adsorption in BET measurements. However, with increasing transformation time the SSA dropped again. Similar observations were made during prolonged ageing of schwertmannite from the Monte Romero abandoned mine (Iberian Pyrite Belt, SW Spain), where schwertmannite completely transformed into goethite and jarosite (Acero et al., 2006).

Thermal exposure of schwertmannite led to a gradual release of sulphate, with the sulphate primarily remaining within the structure at temperatures below 200°C (Henderson and Sullivan, 2010; Johnston et al., 2016). However, at temperatures exceeding 400°C, sulphate migrated from the lattice to the surface from where it is released into solution, and at 600°C, sulphate volatilized largely as SO_2 and SO. Under alkaline conditions (pH > 13), most sulphate is released when schwertmannite is exposed to temperatures ≥60°C, with sulphate release increasing as both ageing temperature and time increase (Davidson et al., 2008). This increased crystallinity of iron oxides likely renders sulphate reduction thermodynamically more favourable than Fe(III) reduction.

6.6.2 Sorbed metal and metalloid ions

In addition to stabilizing schwertmannite against mineralogical transformations, the release of adsorbed trace metals is generally minimal. For example, Paikaray and Peiffer (2012) reported almost no As release after 40 days of ageing of schwertmannite containing 0.92 wt% As(III) at pH 8. Negligible release of As, Pb, Zn, Mn, and Cu has been observed following schwertmannite transformation to goethite under acidic conditions in mine drainage sites such as the Monte Romero mine drainage (SW Spain) (Acero et al., 2005) and the Alta Pb–Ag mine (Montana, USA) (Schroth and Parnell, 2005). The re-adsorption of released As onto newly formed mineral phases may further limit schwertmannite dissolution. Only a minor fraction of As(V) was released during schwertmannite transformation to goethite and lepidocrocite at pH 6.5, with more than 99% remaining within the solid phase (Burton and Johnston, 2012). Arsenic release is negligible even in anoxic environments, as goethite retains a major fraction of As following schwertmannite transformation.

The release of As(V) from As(V)-rich schwertmannite during ageing varies depending on the schwertmannite properties and the pH-dependent speciation of arsenate. At pH 2, where $H_3AsO_4^0$ is the predominant species, a higher As release is observed compared to pH 10, where AsO_4^{3-} dominates (Wang et al., 2020). This difference is attributed to repulsive forces between the positively charged schwertmannite surface and $H_3AsO_4^0$, whereas AsO_4^{3-} readily exchanges with OH^- and SO_4^{2-} ions (Wang et al., 2020). In contrast, when As(V) is co-precipitated with schwertmannite during synthesis, an opposite trend is observed. In this case, schwertmannite containing co-precipitated As(V) exhibits a smaller particle size and greater SSA, which limits As(V) release at pH 10 compared to pH 2 due to the neutral charge of As at higher pH levels (Wang et al., 2020).

Nonetheless, As remobilization can increase over time as mineral transformation progresses. While As concentrations remain low in environments dominated by schwertmannite, the predominance of goethite and the gradual disappearance of schwertmannite accompanied by drop in pH from ~3 to ~1.7 can trigger As release due to the lower sorptive capacity of goethite (Acero et al., 2005, 2006). Despite this, the extent of As release remains minimal, accounting for only 0.06% (Acero et al., 2005) or 0.02% (Paikaray and Peiffer, 2015) of the total As content at pH ~ 3 and ~8, respectively, indicating that most As is incorporated into newly formed, more crystalline precipitates. A similar increase in As release has been observed upon thermal exposure of As-rich schwertmannites at elevated temperatures (Johnston et al., 2016).

The presence of competitive compounds, such as silicic acid, can enhance As release during schwertmannite transformation. Orthosilicate (H_4SiO_4 and $H_3SiO_4^-$) and polymeric Si species exhibit a strong affinity for Fe(III) minerals and compete with As for sorption sites (Swedlund and Webster, 1999; Luxton et al., 2006, 2008). Consequently, As concentration increased from 0.1 µM in the absence of dissolved Si to 0.27 µM at a Si concentration of 9.5 mM at pH 6.5 (Burton and Johnston, 2012).

Arsenic release is minimized in the presence of DIRB, such as *G. sulfurreducens*, compared to controls lacking bacterial cells (Cutting et al., 2012). This is attributed to the acceleration of schwertmannite transformation by microbial activity, a process that allows for re-adsorption of the released As. Notably, solid-phase As(V) does not appear to be reduced to As(III) during schwertmannite reduction (Burton et al., 2010; Fan et al., 2019), whereas Cr(VI) undergoes partial reduction to Cr(III) (Fan et al., 2019).

The tendency of heavy metal ions to become mobilized decreases with their affinity for adsorption onto schwertmannite (Li et al., 2018). The release of trace metals coincides with schwertmannite dissolution, although many of these metals become re-adsorbed onto newly formed mineral phases, as observed for Cu^{2+} (Antelo et al., 2013). However, Pb^{2+} retention increases over time due to Pb–SO_4 complexation on goethite surfaces (Acero et al., 2005, 2006).

6.7 Summary

Schwertmannite is widely distributed around AMD, natural streams, pit lakes, and ASS localities. The SO_4^{2-} content in schwertmannite usually ranges between 0.1 and 15.5 wt% for synthetic and between 8.5 and 14.1 wt% for natural specimens, while the iron content varies between 43.2 and 56.7 wt% and between 39.4 and 41.8 wt%, respectively.

Schwertmannite is a metastable iron sulphate mineral prevalent in environments impacted by AMD and ASSs. Its transformation into more stable iron phases like goethite, jarosite, ferrihydrite, or hematite is influenced by environmental conditions, especially pH, redox state, and the presence of other ions.

Schwertmannite transforms primarily through dissolution–reprecipitation processes, significantly influenced by pH. At acidic pH (pH ~ 3), the transformation to goethite is slow and can take months to years. However, higher pH accelerates this transformation, often within days. The process releases protons, making it acid-generating, and is accompanied by a drop in pH. Schwertmannite stabilizes AMD environments by buffering pH at ~3.

At highly acidic conditions (pH < 2.5), jarosite and goethite dominate. Cations like K^+, NH_4^+, and H_3O^+ promote jarosite formation. The presence of bacteria, such as *A. ferrooxidans*, can also influence these transformations.

Schwertmannite's poor crystallinity and high surface area allow it to adsorb contaminants like arsenic, phosphate, and metals. These adsorbed or incorporated ions stabilize the mineral and inhibit transformation by reducing Fe^{3+} and SO_4^{2-} release and preventing nucleation of new phases. High phosphate loading delays goethite formation significantly, and similarly, high As(V) content can halt transformation under extreme pH conditions. NOM, structural sulphate content, and high concentrations of dissolved SO_4^{2-} also inhibit transformation by limiting Fe^{3+} activity or passivating the mineral surface.

In oxygen-limited environments, such as sediment layers in AMD lakes or wetlands, microbial reduction processes dominate. Schwertmannite undergoes reductive dissolution by DIRB and SRBs, producing Fe(II), bicarbonate, and hydrogen sulphide, shifting pH towards neutral or alkaline. This leads to formation of secondary minerals like goethite, mackinawite (FeS), siderite ($FeCO_3$), and even magnetite under Fe(II)-rich conditions. Reduction is enhanced by microbes such as *G. sulfurreducens*, particularly when Fe(II) is available to catalyze electron transfer, destabilizing schwertmannite. Under these conditions, both solid-state and dissolution-based transformation pathways occur.

Fe(II) in solution catalyzes schwertmannite transformation, especially at pH > 5. It adsorbs onto schwertmannite surfaces and transfers electrons to Fe(III), initiating breakdown and recrystallization into goethite. The transformation is rapid, reducing timescales from weeks to hours under optimal $Fe(II)_{aq}$ and pH conditions. However, high concentrations of coexisting anions like chromate (CrO_4^{2-}), phosphate, and NOM inhibit Fe(II) adsorption, thus slowing transformation. Silica, particularly at circum-neutral to alkaline pH, also adsorbs onto schwertmannite surfaces, blocking Fe(II) access and hindering structural changes.

Schwertmannite remains stable below 200°C. At 400°C and above, it begins to transform into goethite and hematite, with hematite becoming dominant beyond 600°C. OM accelerates this process by serving as a reducing agent. In OM-rich environments, intermediate phases such as ferrihydrite, siderite, mackinawite, and magnetite appear during heating, while arsenic bound to schwertmannite may be reduced from As(V) to As(III) during thermal treatment. Transformation pathways at elevated temperatures involve dehydration, structural sulphate release, and solid-state transitions. Ultimately, hematite emerges as the final product under high heat, particularly in fire-affected landscapes.

Schwertmannite transformation involves release of Fe^{3+} and SO_4^{2-}, particularly during dissolution stages. While Fe^{3+} remains low due to re-precipitation, sulphate is released through ligand exchange and structural breakdown. The extent of release depends on pH, ionic strength, temperature, and presence of competing ions or contaminants. Trace metals (e.g., As, Cu, Pb, Zn) are typically retained or re-adsorbed during transformation. However, at extreme pH or elevated temperatures, minor release may occur, especially when sorption sites are disrupted.

References

Acero P, Ayora C, Torrentó C and Nieto J (2006). The behavior of trace elements during schwertmannite precipitation and subsequent transformation into goethite and jarosite. *Geochimica et Cosmochimica Acta* 70: 4130–4139.

Acero P, Torrentó C and Ayora C (2005). Effect of schwertmannite ageing on acid rock drainage geochemistry. In: *9th international mine water congress*, IMWA Springer-Verlag, Spain, pp. 67–73.

Antelo J, Fiol S, Gondar D, Pérez C, López R and Arce F (2013). Cu(II) incorporation to schwertmannite: Effect on stability and reactivity under AMD conditions. *Geochimica et Cosmochimica Acta* 119: 149–163.

As K, Peiffer S, Uhuegbue PO, Joshi P, Kappler A, Marouane B and Hockmann K (2024) Sulfate affinity controls phosphate sorption and the proto-transformation of schwertmannite. *Chemical Geology* 653: 122043.

Barham BJ (1997). Schwertmannite: A unique mineral, contains a replaceable ligand, transforms to jarosites, hematites, and/or basic iron sulfate. *Journal of Material Research* 12: 2751–2757.

Bertel D, Peck J, Quick TJ and Senko JM (2012). Iron transformations induced by an acid-tolerant *Desulfosporosinus* species. *Applied Environmental Microbiology* 78: 81–88.

Bigham JM, Carlson L and Murad E (1994). Schwertmannite, A new iron oxyhydroxysulphate from Pyhäsalmi, Finland, and other localities. *Mineralogical Magazine* 58: 641–648.

Bigham JM, Schwertmann U and Carlson L (1992). Mineralogy of precipitates formed by the biogeochemical oxidation of Fe(II) in mine drainage. In: Skinner HGW and Fitzpatrick RW (Eds), *Biomineralization processes of iron and manganese—Modern and ancient environments.* Catena Supplement Catena-Verlag, pp. 219–232.

Bigham JM, Schwertmann U, Carlson L and Murad E (1990). A poorly crystallized oxyhydroxysulfate of iron formed by the bacterial oxidation of Fe(II) in acid mine waters. *Geochimica et Cosmochimica Acta* 54: 2743–2758.

Bigham JM, Schwertmann U and Pfab G (1996b). Influence of pH on mineral speciation in a bioreactor stimulating acid mine drainage. *Applied Geochemistry* 11: 845–849.

Bigham JM, Schwertmann U, Traina SJ, Winland RL and Wolf M (1996a). Schwertmannite and the chemical modeling of iron in acid sulfate waters. *Geochimica et Cosmochimica Acta* 60: 2111–2121.

Blake D, Lu K, Horwitz P and Boyce MC (2012). Fire suppression and burnt sediments: Effects on the water chemistry of fire-affected wetlands. *International Journal of Wildland Fire* 21: 557–561.

Blodau C (2004). Evidence for a hydrologically controlled iron cycle in acidic and iron rich sediments. *Aquatic Science* 66: 47–59.

Blodau C (2006). A review of acidity generation and consumption in acidic coal mine lakes and their watersheds. *Science of the Total Environment* 369: 307–332.

Bradstock RA, Bedward M and Cohn JS (2006). The modelled effects of differing fire management strategies on the conifer *Calltris verrucosa* within semi-arid mallee vegetation in Australia. *Journal of Applied Ecology* 43: 281–292.

Brady KS, Bigham JM, Jaynes WF and Logan TJ (1986). Influence of sulfate on Fe-oxide formation: Comparisons with a stream receiving acid mine drainage. *Clays and Clay Minerals* 34: 266–274.

Burton ED, Bush RT and Sullivan LA (2006). Sedimentary iron geochemistry in acidic waterways associated with coastal lowland acid sulfate soils. *Geochimica et Cosmochimica Acta* 70: 5445–5468.

Burton ED, Bush RT, Sullivan LA and Mitchell DRG (2007). Reductive transformation of iron and sulfur in schwertmannite-rich accumulations associated with acidified coastal lowlands. *Geochimica et Cosmochimica Acta* 71: 4456–4473.

Burton ED, Bush RT, Sullivan LA and Mitchell DRG (2008b). Schwertmannite transformation to goethite via the Fe(II) pathway: Reaction rates and implications for iron-sulfide formation. *Geochimica et Cosmochimica Acta* 72: 4551–4564.

Burton ED, Bush RT, Sullivan LA, Johnston SG and Hocking RK (2008a). Mobility of arsenic and selected metals during re-flooding of iron- and organic-rich acid-sulfate soil. *Chemical Geology* 253: 64–73.

Burton ED and Johnston SG (2012). Impact of silica on the reductive transformation of schwertmannite and the mobilization of arsenic. *Geochimica et Cosmochimica Acta* 96: 134–153.

Burton ED, Johnston SG, Kraal P, Bush RT and Claff S (2013). Sulfate availability drives divergent evolution of arsenic speciation during microbially mediated reductive transformation of schwertmannite. *Environmental Science & Technology* 47: 2221–2229.

Burton ED, Johnston SG, Watling K, Bush RT, Keene AF and Sullivan LA (2010). Arsenic effects and behavior in association with the Fe(II)-catalyzed transformation of schwertmannite. *Environmental Science & Technology* 44: 2016–2021.

Carlson L, Bigham JM, Schwertmann U, Kyek A and Wagner F (2002). Scavenging of As from acid mine drainage by schwertmannite and ferrihydrite: a comparison with synthetic analogues. *Environmental Science & Technology* 36: 1712–1719.

Chen Q, Cohen DR, Andersen MS, Robertson AM and Jones DR (2022). Stability and trace element composition of natural schwertmannite precipitated from acid mine drainage. *Applied Geochemistry* 143: 105370.

Choppala G and Burton ED (2018). Chromium(III) substitution inhibits the Fe(II)-accelerated transformation of schwertmannite. *PLOS ONE* 13: e0208355.

Collins RN, Jones AM and Waite TD (2010). Schwertmannite stability in acidified coastal environments. *Geochimica et Cosmochimica Acta* 74: 482–496.

Cornell RM and Schwertmann U (2003). *The iron oxides: Structure, properties, reactions, occurrences and uses.* Wiley-VCH.

Courtin-Nomade A, Grosbois C, Bril H and Roussel C (2005). Spatial variability of arsenic in some iron-rich deposits generated by acid mine drainage. *Applied Geochemistry* 20: 383–396.

Cruz-Hermández P, Peréz-López R and Nieto JM (2017). Role of arsenic during the aging of acid mine drainage precipitates. *Proceedia Earth and Planetary Science* 17: 233–236.

Cudennec Y and Lecerf A (2005). Topotactic transformations of goethite and lepidocrocite into hematite and maghemite. *Solid State Science* 7: 520–529.

Cutting RS, Coker VS, Telling ND, Kimber RL, Laan G, Pattrick RAD, Vaughan DJ, Arenholz E and Lloyd JR (2012). Microbial reduction of arsenic-doped schwertmannite by *Geobacter sulfurreducens. Environmental Science & Technology* 46: 12591–12599.

Davidson LE, Shaw S and Benning LG (2008). The kinetics and mechanism of schwertmannite transformation to goethite and hematite under alkaline conditions. *American Mineralogists* 93: 1326–1337.

Doelsch E, Rose J, Masion A, Bottero JY, Nohon D and Bertsch PM (2000). Speciation and crystal chemistry of iron(III) chloride hydrolysed in the presence of SiO_4 ligands. An Fe K-edge EXAFS study. *Langmuir* 16: 4726–4731.

Fan C, Guo C, Chen M, Huang W, Wan J, Reinfelder JR, Li X, Zeng Y, Lu X and Dang Z (2018). Transformation of cadmium-associated schwertmannite and subsequent element repartitioning behaviors. *Environmental Science and Pollution Research* 26: 617–627.

Fan C, Guo C, Zeng Y, Tu Z, Ji Y, Reinfelder JR, Chen M, Huang W, Lu G, Yi X and Dang Z (2019). The behavior of chromium and arsenic associated with redox transformation of schwertmannite in AMD environment. *Chemosphere* 222: 945–953.

Fukushi K, Sasaki M, Sato T, Yanase N, Amano H and Ikeda H (2003). A natural attenuation of arsenic in drainage from an abandoned arsenic mine dump. *Applied Geochemistry* 18: 1267–1278.

Fukushi K, Sato T, Yanase N, Minato J and Yamada H (2004). Arsenate sorption on schwertmannite. *American Mineralogists* 89: 1728–1734.

Gagliano WB, Brill MR, Bigham JM, Jones FS and Traina SJ (2004). Chemistry and mineralogy of ochreous sediments in a constructed mine drainage wetland. *Geochimica et Cosmochimica Acta* 68, 2119–2128.

Gorski CA and Scherer MM. (2011). Fe^{2+} sorption at the Fe oxide-water interface: A revised conceptual framework. In: Tratnyek, PG, Grundl, TJ und Haderlein, SB (Hg.), *Aquatic redox chemistry.* American Chemical Society (ACS Symposium Series), S. 315–343.

Henderson SP and Sullivan LA (2010). Low temperature transformation of schwertmannite to hematite with associated CO_2, SO and SO_2 evolution. In: Gilkes R, Prakongkep N, Gilkes R. (Ed.), *19th world congress soil science and soil solutions for a changing world*, Brisbane, Australia, pp. 72–75.

Henderson SP, Sullivan LA, Bush RT and Burton ED (2007). Schwertmannite transformation to hematite by heating: Implications for pedogenesis, water quality and CO_2/SO_2 export in acid sulfate soil landscapes. *Geochimica et Cosmochimica Acta* 71: A394.

Henderson SP, Sullivan LA, Bush RT and Burton ED (2008). Thermal transformation of schwertmannite to hematite: anomalous stored acidity. In: Lin C, Huang S and Li Y (Eds), *Proceedings of joint conference of 6th international acid sulfate soil conference acid rock drainage.* Guangdong Press Group, Guangzhou, p. 264.

Hockridge JG, Jones F, Loan M and Richmond WR (2009). An electron microscopy study of the crystal growth of schwertmannite needles through oriented aggregation of goethite nanocrystals. *Journal of Crystal Growth* 311: 3876–3882.

Houngaloune S, Hiroyoshi N and Ito M (2015a). Effect of Fe(II) and Cu(II) on the transformation of schwertmannite to goethite under acidic condition. *International Journal of Chemical Engineering and Applications* 6: 32–37.

Houngaloune S, Hiroyoshi N and Ito M (2015b). Stability of As(V)-sorbed schwertmannite under porphyry copper mine conditions. *Minerals Engineering* 74: 51–59.

Janneck E, Burghardt D, Martin M, Damian C, Schöne G, Meyer J and Peiffer S (2011). From waste to valuable substance: Utilization of schwertmannite and lignite filter ash for removal of arsenic and uranium from mine drainage. In: Rüde T, Freund A and Wolkersdorfer C (Eds), *Mine water-managing the challenges & IMWA 2011*, IMWA, Aachen, Germany, pp. 359–364.

Janneck E, Burghardt D, Simon E, Peiffer S, Paul M and Koch T (2015). Development of an adsorbent comprising schwertmannite and its utilization in mine water treatment. In: Brown A, Bucknam C, Burgess J, Carballo M, Castendyk D, Figueroa L, Kirk L, McLemore V, McPhee J, O'Kane M, Seal R, Wiertz J, Williams D, Wilson W, Wolkersdorfer C. (Eds), *10th international conference on acid rock drainage & IMWA annual conference 2015*, Santiago, Chile, pp. 1–10.

Johnston SG, Bennett WW, Burton ED, Hockmann K, Dawson N and Karimian N (2018). Rapid arsenic(V)-reduction by fire in schwertmannite-rich soil enhances arsenic mobilization. *Geochimica et Cosmochimica Acta* 227: 1–18.

Johnston SG, Burton ED, Keene AF, Planer-Friedrich B, Voegelin A, Blackford MG and Lumpkin GR (2012). Arsenic mobilization and iron transformations during sulfidization of As(V)-bearing jarosite. *Chemical Geology* 334: 9–24.

Johnston SG, Burton ED and Moon EM (2016). Arsenic mobilization is enhanced by thermal transformation of schwertmannite. *Environmental Science &Technology* 50: 8010–8019.

Jones AM, Collins RN, Rose J and Waite TD (2009). The effect of silica and natural organic matter on the Fe(II)-catalyzed transformation and reactivity of Fe(III) minerals. *Geochimica et Cosmochimica Acta* 73: 4409–4422.

Jones EJP, Nadeau T, Voytek MA and Landa ER (2006). Role of microbial iron reduction in the dissolution of iron hydroxysulfate minerals. *Journal of Geophysical Research* 111: 1–8.

Jönsson J, Jönsson J and Lövgren L (2006). Precipitation of secondary Fe(III) minerals from acid mine drainage. *Applied Geochemistry* 21: 437–445.

Jönsson J, Persson P, Sjoberg S and Lovgren L (2005). Schwertmannite precipitated from acid mine drainage: Phase transformation, sulphate release and surface properties. *Applied Geochemistry* 20: 179–191.

Jozwiak WK, Kaczmarek E, Maniecki TP, Ignaczak W and Maniukiewicz W (2007). Reduction behavior of iron oxides in hydrogen and carbon monoxide atmospheres. *Applied Catalysis A: General* 326: 17–27.

Ke C, Deng Y, Zhang S, Ren M, Liu B, He J, Wu R, Dang Z and Guo C (2024). Sulfate availability drives the reductive transformation of schwertmannite by co-cultured iron- and sulfate-reducing bacteria. *Science of the Total Environment* 906:167690.

Khamphila K, Kodama R, Sato T and Otake T (2017). Adsorption and post adsorption behavior of schwertmannite with various oxyanions. *Journal of Minerals and Materials Characterization and Engineering* 5: 90–106.

Kim H and Kim Y (2021). Schwertmannite transformation to goethite and related mobility of trace metals in acid mine drainage. *Chemosphere* 269: 128720.

Klug MA, Janneck E, Reichel S and Peiffer S (2016). Treatment of chromate(VI) and vanadate(V) polluted wastewaters using schwertmannite adsorbents. In: Drebenstedt C and Paul M (Eds), *Mining meets water - Conflicts and solutions & IMWA 2016*, IMWA, Leipzig, Germany, pp. 1012–1014.

Knorr K and Blodau C (2007). Controls on schwertmannite transformation rates and products. *Applied Geochemistry* 22: 2006–2015.

Kumpulainen S, Carlson L and Raisanen ML (2007). Seasonal variations of ochreous precipitates in mine effluents in Finland. *Applied Geochemistry* 22: 760–777.

Kumpulainen S, Räisänen ML, Von der Kammer F and Hofmann T (2008). Ageing of synthetic and natural schwertmannites at pH 2–8. *Clay Minerals* 43: 437–448.

Li J, Xie Y, Lu G, Ye H, Yi X, Reinfelder JR, Lin Z and Dang Z (2018). Effect of Cu(II) on the stability of oxyanion-substituted schwertmannite. *Environmental Science and Pollution Research* 25: 15492–15506.

Liao Y, Liang J and Zhou L (2011). Adsorptive removal of As(III) by biogenic schwertmannite from simulated As-contaminated groundwater. *Chemosphere* 83: 295–301.

Luxton TP, Eick MJ and Rimstidt DJ (2008). The role of silicate in the adsorption/desorption of arsenite on goethite. *Chemical Geology* 252: 125–135.

Luxton TP, Tadanier CJ and Eick MJ (2006). Mobilization of arsenite by competitive interaction with silicic acid. *Soil Science Society of America Journal* 70: 204–214.

Mazzetti L and Thistlethwaite PJ (2002). Raman spectra and thermal transformations of ferrihydrite and schwertmannite. *Journal of Raman Spectroscopy* 33: 104–111.

Murad E and Rojik P (2003). Iron-rich precipitates in a mine drainage environment: Influence of pH on mineralogy. *American Mineralogists* 88: 1915–1918.

Murad E and Rojik P (2004). Jarosite, schwertmannite, goethite, ferrihydrite and lepidocrocite: The legacy of coal and sulfide ore mining. In: Singh B (Ed), *3rd Australian New Zealand soils conference, SuperSoil 2004,* Sydney, Australia, pp. 1–8.

Paikaray S and Peiffer S (2010). Dissolution kinetics of sulfate from schwertmannite under variable pH conditions. *Mine Water and the Environment* 29: 263–269.

Paikaray S and Peiffer S (2012). Abiotic schwertmannite transformation kinetics and the role of sorbed As(III). *Applied Geochemistry* 27: 590–597.

Paikaray S and Peiffer S (2015). Lepidocrocite formation kinetics from schwertmannite in Fe(II)-rich anoxic alkaline medium. *Mine Water and the Environment* 34: 213–222.

Paikaray S, Essilfie-Dughan J, Göttlicher J, Pollock K and Peiffer S (2014). Redox stability of As(III) on schwertmannite surfaces. *Journal of Hazardous Materials* 265: 208–216.

Paikaray S, Göttlicher J and Peiffer S (2011). Removal of As(III) from acidic waters using schwertmannite: Surface speciation and effect of synthesis pathway. *Chemical Geology* 283: 134–142.

Paikaray S, Göttlicher J and Peiffer S (2012). As(III) retention kinetics, equilibrium and redox stability on biosynthesized schwertmannite and its fate and control on schwertmannite stability on acidic (pH 3.0) aqueous exposure. *Chemosphere* 86: 557–564.

Paikaray S, Schröder C and Peiffer S (2017). Schwertmannite stability in anoxic Fe(II)-rich aqueous solution. *Geochimica et Cosmochimica Acta* 217: 292–305.

Parfitt RL and Smart RC (1978). The mechanism of sulfate adsorption on iron oxides. *Soil Science Society of America Journal* 42: 48–50.

Pállová Z, Kupka D and Achimovičová M (2010). Metal mobilization from AMD sediments in connection with bacterial iron reduction. *Mineralia Slovaca* 42: 343–347.

Pedersen HD, Postma D, Jakobsen R and Larsen O (2005). Fast transformation of iron oxyhydroxides by the catalytic action of aqueous Fe(II). *Geochimica et Cosmochimica Acta* 69: 3967–3977.

Peine A, Küsel K, Tritschler A and Peiffer S (2000). Electron flow in an iron-rich acidic sedimant - Evidence for an acidity-driven iron cycle. *Limnology and Oceanography* 45: 1077–1087.

Peretyazhko T, Zachara JM, Boily JF, Xia Y, Gassman PL, Arey BW and Burgos WD (2009). Mineralogical transformations controlling acid mine drainage chemistry. *Chemical Geology* 262: 169–178.

Qiao X, Liu L, Shi J, Zhou L, Guo Y, Ge Y, Fan W and Liu F (2017). Heating changes bio-schwertmannite microstructure and arsenic(III) removal efficiency. *Minerals* 7: 1–14.

Randall SR, Sherman DM, Ragnarsdottir KV and Collins CR (1999). The mechanism of cadmium surface complexation on iron oxyhydroxide minerals. *Geochimica et Cosmochimica Acta* 63: 2971–2987.

Regenspurg S, Brand A and Peiffer S (2004). Formation and stability of schwertmannite in acidic mining lakes. *Geochimica et Cosmochimica Acta* 68: 1185–1197.

Regenspurg S and Peiffer S (2005). Arsenate and chromate incorporation in schwertmannite. *Applied Geochemistry* 20: 1226–1239.

Reichelt L and Bertau M (2015). Transformation of nanostructured schwertmannite and 2-Line-ferrihydrite into hematite. *Journal of Inorganic and General Chemistry* 641: 1696–1700.

Sánchez-España J, Yusta I and López GA (2012). Schwertmannite to jarosite conversion in the water column of an acidic mine pit lake. *Mineralogical Magazine* 76: 2659–2682.

Schoepfer VA, Burton ED, Johnston SG and Kraal P (2017). Phosphate-imposed constraints on schwertmannite stability under reducing conditions. *Environmental Science & Technology* 51: 9739–9746.

Schoepfer VA, Burton ED and Johnston SG (2019). Contrasting effects of phosphate on the rapid transformation of schwertmannite to Fe(III) (oxy)hydroxides at near-neutral pH. *Geoderma* 340: 115–123.

Schonberger S (2016). Stability of schwertmannite and cobalt substituted schwertmannite in mining environments. In: Davidson C and Wirth K (Eds), *29th annual symposium on learning science through research*, Keck Symposium Volume, Oberlin College, Ohio, USA, pp. 1–5, ISBN: 1528–7491.

Schroth AW and Parnell RA (2005). Trace metal retention through the schwertmannite to goethite transformation as observed in a field setting, Alta Mine, MT. *Applied Geochemistry* 20: 907–917.

Schwertmann U and Carlson L (2005). The pH-dependent transformation of schwertmannite to goethite at 25°C. *Clay Minerals* 40: 63–66.

Schwertmann U and Murad E (1983). Effect of pH on the formation of goethite and hematite from ferrihydrite. *Clays and Clay Minerals* 31: 277–284.

Schwertmann U, Bigham JM and Murad E (1995). The first occurrence of schwertmannite in a natural stream environment. *European Journal of Mineralogy* 7: 547–552.

Sullivan LA and Bush RT (2004). Iron precipitate accumulations associated with waterways in drained coastal acid sulfate landscapes of eastern Australia. *Marine and Freshwater Research* 55: 727–736.

Swedlund PJ and Webster JG (1999). Adsorption and polymerization of silicic acid on ferrihydrite, and its effect on arsenic adsorption. *Water Research* 33: 3413–3422.

Swedlund PJ and Webster JG (2001). Cu and Zn ternary surface complex formation with SO_4 on ferrihydrite and schwertmannite. *Applied Geochemistry* 16: 503–511.

Vithana CL, Johnston SG and Dawson N (2018). Divergent repartitioning of copper, antimony and phosphorus following thermal transformation of schwertmannite and ferrihydrite. *Chemical Geology* 483: 530–543.

Vithana CL, Sullivan LA, Burton ED and Bush RT (2015). Stability of schwertmannite and jarosite in an acidic landscape: Prolonged field incubation. *Geoderma* 239–240: 47–57.

Walter M, Arnold T, Reich T and Bernhard G (2003). Sorption of uranium(VI) onto ferric oxides in sulfate-rich acid waters. *Environmental Science & Technology* 37: 2898–2904.

Wang H, Bigham JM and Tuovinen OH (2006). Formation of schwertmannite and its transformation to jarosite in the presence of acidophilic iron-oxidizing microorganisms. *Material Science and Engineering C* 26: 588–592.

Wang Y, Gao M, Huang W, Wang T and Liu Y (2020). Effects of extreme pH conditions on the stability of As(V)-bearing schwertmannite. *Chemosphere* 251: 126427.

Wang X, Gu C, Feng X and Zhu M (2015). Sulfate local coordination environment in schwertmannite. *Environmental Science & Technology* 49: 10440–10448.

Waychunas GA, Xu N, Fuller CC, Davis JA and Bigham JM (1995). XAS study of AsO_4^{3-} and SeO_4^{2-} substituted schwertmannites. *Physica B*, 208–209: 481–483.

Webster JG, Swedlund PJ and Webster KS (1998). Trace metal adsorption onto a acid mine drainage iron(III) oxy hydroxy sulfate. *Environmental Science & Technology* 32: 1361–1368.

Winland RL, Traina SJ and Bigham JM (1991). Chemical composition of ochreous precipitates from Ohio coal mine drainage. *Journal of Environmental Quality* 20: 452–460.

Ying H, Feng X, Zhu M, Lanson B, Liua F and Wang X (2020). Formation and transformation of schwertmannite through direct Fe^{3+} hydrolysis under various geochemical conditions. *Environmental Science Nano* 7, 2385.

Yu JY, Park M and Kim J (2002). Solubilities of synthetic schwertmannite and ferrihydrite. *Geochemical Journal* 36: 119–132.

Zhang Y, Gao K, Dang Z, Huang W, Reinfelder JR and Ren Y (2021). Microbial reduction of As(V)-loaded schwertmannite by *Desulfosporosinus meridiei*. *Science of the Total Environment* 764: 144279.

INDEX

For Product Safety Concerns and Information please contact our EU representative GPSR@taylorandfrancis.com
Taylor & Francis Verlag GmbH, Kaufingerstraße 24, 80331 München, Germany

www.ingramcontent.com/pod-product-compliance
Lightning Source LLC
LaVergne TN
LVHW020712110826
845149LV00012B/2225
* 9 7 8 1 0 3 2 9 5 2 3 8 3 *